Biotechnology in the Food Industry

M. P. Tombs

Open University Press
Milton Keynes

Open University Press
Celtic Court
22 Ballmoor
Buckingham MK18 1XW

First Published 1990

Copyright © M.P. Tombs 1990

All rights reserved. No part of this publication may be
reproduced, stored in a retrieval system or transmitted in
any form or by any means, without written permission from the
publisher.

British Library Cataloguing in Publication Data

Tombs, M. P.
 Biotechnology in the food industry.
 1. Food. technology. Applications of biotechnology
 I. Title II. Series
 644

 ISBN 0-335-15899-4
 ISBN 0-335-15898-6 (pbk)

ANDERSONIAN LIBRARY

0 1. MAR 91

UNIVERSITY OF STRATHCLYDE

Typeset by Vision Typesetting, Manchester
Printed in Great Britain by Biddles Limited, Guildford and King's Lynn

Biotechnology in the Food Industry

The Biotechnology Series

This series is designed to give undergraduates, graduates and practising scientists access to the many related disciplines in this fast developing area. It provides understanding both of the basic principles and of the industrial applications of biotechnology. By covering individual subjects in separate volumes a thorough and straightforward introduction to each field is provided for people of differing backgrounds.

Titles in the Series

Biotechnology: The Biological Principles: M.D. Trevan, S. Boffey, K.H. Goulding and P. Stanbury
Fermentation Kinetics and Modelling: C.G. Sinclair and B. Kristiansen (Ed. J.D. Bu'lock)
Enzyme Technology: P. Gacesa and J. Hubble
Animal Cell Technology: M. Butler
Fermentation Biotechnology: O.P. Ward
Genetic Transformation in Plants: R. Walden
Plant Biotechnology in Agriculture: K. Lindsey and M.G.K. Jones
Biosensors: E. Hall
Biotechnology of Biomass Conversion: M. Wayman and S. Parekh
Biotechnology in the Food Industry: M.P. Tombs

Upcoming Titles

Biotechnology Safety   Applied Gene Technology
Plant Cell and Tissue Culture   Bioreactors

Series Editors

Professor J.A. Bryant   *Department of Biology, Exeter University, England*
Professor J.F. Kennedy   *Department of Chemistry, University of Birmingham, England*

Series Advisers

Professor C.H. Self   *University of Newcastle upon Tyne, England*
Dr R.N. Greenshields   *GB Biotechnology Ltd., Swansea, Wales*

The Institute of Biology IoB

This series has been editorially approved by the **Institute of Biology** in London. The Institute is the professional body representing biologists in the UK. It sets standards, promotes education and training, conducts examinations, organizes local and national meetings, and publishes the journals **Biologist** and **Journal of Biological Education**.

For details about Institute membership write to: Institute of Biology, 20 Queensberry Place, London SW7 2DZ.

Contents

Preface vii
Acknowledgements xi

CHAPTER 1 — **The Basic Operations of Biotechnology** — 1
Making a Start: Basic Principles — 1
Enzyme Nomenclature — 3
Time and Cost — 5
Modification of Lipid Sources — 6
Conclusions — 46

CHAPTER 2 — **Sweeteners** — 47
Introduction — 47
Sucrose — 48
Glucose and Fructose Syrups — 53
Starch Degradation — 53
Glucose from Other Sources — 55
Glucose Isomerization — 55
Immobilized Enzyme Reactors — 58
The Sweet Proteins — 61
Aspartame — 66
Steviosides and Rebaudiosides — 68
Conclusions — 68

CHAPTER 3 — **Mainly Proteins: Proteases, Gels and Fermented Foods** — 71
Introduction: Food Texture — 71
Protein–Protein Interactions and Structure Formation — 72

	Stability	79
	Protein Gelation	82
	Some Fermented Foods	89
	Proteases	100
	Other Covalent Modifications	109
	Conclusions	110
CHAPTER 4	**Cereals: Baking and Brewing**	**112**
	Introduction	112
	Baking	114
	Brewing Beer and Vinegar	121
CHAPTER 5	**Lipases and the Minor Components, Emulsifiers, Stabilizers and Flavours**	**127**
	Introduction	127
	Properties of Lipases	128
	Emulsifiers	134
	Stabilizers and Gelling Agents	137
	Flavours and Enhancers	142
	The Use of Plant-Cell Cultures	143
	Conclusions	146
CHAPTER 6	**Food, Populations and Quality Control**	**147**
	Introduction	147
	Testing and Hazards	148
	Toxic Components of Plants	156
	Population Polymorphisms and Food	166
	General Conclusions and Future Prospects	173

Further Reading   177
Index   182

Preface

This book is intended as an introductory text for third-year undergraduates, and people engaged in research and development in the food industry who are not directly involved in biotechnology research. Thus it assumes that readers have followed a course of chemistry or biochemistry and have at least some acquaintance with the chemistry of foods and naturally occurring materials.

There is no neatly defined academic package called Biotechnology. This does not mean that academics have no interest in it. Most of the advances in molecular biology that have created the field have come from academics. It does mean however that students, who tend to like a clearly defined subject, and above all syllabus, find it difficult to get to grips with the rather more diffuse matter of an interdisciplinary subject. There is not much point in trying to define it. Many of those engaged in research and development in the food industry, if asked, would reply that they were indeed engaged in biotechnology research, and this book is about what they are doing. It is because it has little or no intellectual coherence or academic structure that I have chosen to approach biotechnology from the industrial and applied standpoint. The structure, if structure there is, is that of the industry. In so doing I have nevertheless tried to be academically complete, in dealing in a reasonably rounded manner with the particular areas of biochemistry and biology that are important. There are no references in the text, and I have tried to adopt a narrative style. Above all I have tried to give an impression of work in progress, since so much biotechnology in the food industry has not yet come to fruition.

A chapter-by-chapter reading list of books, reviews and a few original papers is given at the end of the book, and references also accompany the figures. The reviews are chosen for breadth, rather than as the most recent survey. Biotechnology has produced a most diverse literature, ranging from erudite journals to stock-market tip sheets. Very little of it is to do with applications in the food industry. The primary literature in this field is scattered through journals dealing with food science, with microbiology, and explicitly with biotechnology, and much of it has not emerged from patents. Annual reviews in the field of food science, and series such as *Biotechnology and Genetic Engineering Reviews* have some articles on food applications though the latter tends to give more prominence to agriculture.

It is necessary to say a few words about the food industry. It is usually held that it begins at the farm gate, and ends on the consumer's table. It might equally begin at the quay-side as the trawlers come in, but broadly speaking it does not include agriculture or the primary production of raw materials. On the other hand the impact of biotechnology on agriculture is likely to be very great, and the food industry is exceedingly sensitive to its raw material supplies. So the dividing line is not so clear-cut: I have tended to include those developments that look as if they will lead to the appearance of novel raw materials, but not to include those that may simply lead to an enhanced supply of existing raw materials. Thus, oil seeds containing novel lipids are considered, but cereals with enhanced yields are not.

Chapter 1 deals with basic principles, with the aid of some examples, particularly an attempt to modify lipid synthesis. I should emphasize that the basic principles are not just those of the biochemistry involved but also how projects are chosen and why timescales and costs sometimes rule out what seem to be quite attractive ideas from the purely scientific point of view.

I have also drawn on these examples to illustrate the sheer technical difficulty of so much of this work. Existing texts make biotechnology look altogether too easy. Every research worker knows that some materials are easier to work with than others, and they will usually choose the most favourable case to illustrate a principle. The early work on protein synthesis was done on cells which produced large amounts of a single protein. A remarkably large proportion of protein structure work, in its beginnings, was on proteins that came already in solution, in milk or blood or eggs. In industry the choice of system is determined by commercial imperatives, which take no account of experimental facility. This problem is particularly acute in biotechnology which to be really successful depends on the availability of a great deal of detailed information. Chapter 1 explores what kind of information has to be obtained as a prerequisite to the final flourishes of gene manipulation. Later, proteases are considered, though with an underlying theme of fermented products, and foods dependent on protein-derived structures. Fermented drinks are not considered in detail, since there is little sign of novel developments in their production and the biotechnology is traditional and described elsewhere.

Further chapters consider lipids, in their food context, but also lipases as an example of an isolated enzyme that is actually finding application in food processing. Chapter 2 describes the current state of sweeteners, a triumph of biotechnology already, where there are also new biotechnology initiatives in progress, illustrating the whole range from gene transfer to the use of enzyme reactors.

Preface

There are also some potential applications to bread-making and starches and other carbohydrate-based structure builders.

The sort of biotechnology that emphasizes large-scale microbial fermentations actually has rather little application in the food industry, apart from the ancient traditional yeast fermentations for alcoholic drinks. It is likely to be used for the production of feedstocks for synthetic flavours, and colours or the flavours themselves and there are one or two examples.

Finally, I have no doubt that experts in one field or another will feel that I have left out topics that should have been included. I have had to be selective, and I have chosen from the standpoint of the industry rather than the hopeful biotechnologist. I hope that there are not too many errors or anachronisms, though some are I fear inevitable in a rapidly moving field. I have chosen topics, both because they illustrate the point but also because I feel that they have some relevance to the future. However the food industry in the past has developed in quite unexpected ways and may well do so again.

M.P. Tombs

Acknowledgements

It is a pleasure to acknowledge the immense amount that I have learned about biotechnology from my former and present colleagues. Some of them are outstanding experts on the topics considered here and I have drawn heavily on our shared experience. I must also thank Mr Morris Stubbs who has produced incomparable electron micrographs over the years, some of which appear in this book. I am also indebted to all those who gave permission for me to use figures from their publications. Where appropriate the sources are acknowledged.

Chapter 1

The Basic Operations of Biotechnology

Making a Start: Basic Principles

The collection of novel techniques that together have made the development of biotechnology possible have radically altered the feasibility of translating ideas for new processes and products into reality. This is why they are of importance to industry.

Ideas for research and development can come from many sources, but most of them come from other parts of the same commercial organization. Often they come from the marketing department which thinks that it has seen a gap in the market which could be filled by a product that it is not currently available. Or it may be that a competitor has produced something which must be matched and if possible surpassed as quickly as may be. (It is worth noting in passing that marketing departments rarely come up with proposals for new products that would compete with their existing best lines. From their point of view innovation for the mere chance of improvement in sales is so risky that it would be madness to upset an existing successful situation. And indeed it is true that in the food business a high proportion of new introductions do not succeed.)

Research and development (R&D) expenditure has in some instances the quality of an insurance premium, since a primary function of the department is to foresee the appearance of new and significantly better competing products. Nothing is more devastating for a commercial organization than the sudden and unexpected appearance of radically new and superior competition. A good example is the appearance of synthetic detergents where soaps had previously been unopposed. Some traditional soap companies were taken totally by surprise by this event.

Another source of ideas or demands on research is the production departments. This is more often something to do with raw materials than new processes. As markets move, raw materials become more expensive or sometimes vanish altogether. Ways of using less, or substitutes, must be found. This once happened with the commercial supply of saponins, which almost disappeared. Large photographic emulsions need saponins in their manufacture, as spreading agents, and of course diagnostic X-ray machines need large films, so the matter was potentially quite important. Huge quantities of saponins are thrown away as by-products of oilseed processing, but before anything could be done to develop this source the commercial situation eased. Opportunities may be ephemeral and demand a rapid response.

Production departments in the food industry, for obvious reasons, tend to like processes that are relatively undemanding in their raw material requirements. For example there are significant differences between different varieties of potatoes with respect to processing, and research might be asked to devise a process where the kind of potato does not matter. Or they might be asked if they could produce a quite novel kind of potato, perhaps better adapted to the manufacture of chips. (The English chip is known in the USA as 'French fries'.)

Another source of problems for research to solve is the activities of governmental regulating authorities. They may raise objections to long-standing chemical processes which then need replacement. Enzyme-based processes can often offer considerable advantages in the present climate of opinion, which tends to favour so-called 'natural products'. Enzymes are certainly not unnatural, but more seriously can almost always be used to make products with a range of minor components different from those obtained by chemical processing. This is the real advantage of enzymes, though the public perception of enzymes as 'safe' is also of some consequence.

Industrial structures change and vertical integration, in which processors tend to have control of their own raw material supplies, went out of fashion for commercial reasons, and was replaced by a system in which raw materials were purchased as well as could be arranged in the open market. Clearly, a shift of this kind has considerable implications for the demands that are likely to be made on food research, and the ways in which the opportunities of biotechnology are exploited. A dislocation between raw material suppliers and users is more likely when they have different interests, and questions of agricultural economics, and trade policy can get in the way of otherwise attractive pieces of biotechnology, and lead to plenty of work for the development department.

Another source of ideas for R & D is the research department itself. As suggested above, one of its more important functions is to keep an eye on developments in the scientific world, and there will be times when what seem to be worthwhile opportunities for development are seen. These almost always present an immediate problem, because they will require several years of work before anything comes of them. Companies in the food industry have detailed plans for, at the most, a year ahead. Some of them have rather vaguer five-year rolling plans for their future, and the wisdom of hindsight suggests that few of these are actually adhered to.

The Basic Operations of Biotechnology 3

Biotechnology developments, as will appear below, have an exceptionally long time scale, and bring out the clash of time scales in an acute form. It requires a considerable act of faith for a commercial organization to embark on a research programme which will yield results only in about 15 years.

Finally, ideas may come from independent inventors, who are rare birds indeed, though more numerous in the USA than in Europe, and increasingly from academics. University departments are making it their business nowadays to be aware of commercial interests, and the potential applications of their research. Their proposals are of much the same kind as those originating in the research department, and suffer the same problems of time scale. Their main advantage is likely to be early access to novel information before it becomes generally known.

FEASIBILITY

Wherever ideas originate, research will be asked to give a view on their feasibility. Unfortunately this means different things to different people and is the source of much misunderstanding. The scientist is apt to treat the question as one of technical feasibility: that is to say, can the operation be done at all? For example, there can be little doubt that a proposal to sequence the human genome is feasible. All the necessary techniques are known, and there is no reason to expect any particular difficulties. The average non-technical company director does not mean this at all when he asks about feasibility. He wants to know if it can be done within all the constraints of manpower and cost with which he is primarily concerned. This is so obvious to him that he does not even mention it. The idea of technical feasibility is probably quite alien to him, and so far as he is concerned, a proposal to sequence the human genome is definitely not feasible.

The importance of the collection of techniques that have led to biotechnology is that they have changed the feasibility of doing many attractive things in food production and processing. But they have only changed the technical feasibility, and all the other constraints still exist.

Enzyme Nomenclature

The Enzyme Commission (EC), a body set up by the International Union for Biochemistry, has set up a systematic nomenclature for enzymes consisting of a name and a set of four numbers.

The first runs from 1 to 6, and indicates the general class thus:

(1) Oxidoreductases
(2) Transferases
(3) Hydrolases
(4) Lyases
(5) Isomerases
(6) Ligases.

The second indicates the general type of substrate, e.g. nucleic acid or carbohydrate. The third describes a specific coenzyme or substrate requirement. The fourth is the serial number of the enzyme itself in the general list. Obviously there are many millions of different enzymes, but few have been described in sufficient detail to be listed, and the total so far is less than 10 000.

Although the great majority of the enzymes mentioned in this book are hydrolases, there are some others in use in food processing. It is important where possible to mention the EC number, since enzymes even with similar function are often sufficiently different to give practical problems if the right one is not used. Thus, where it has been established, the EC number will be given on first mention of an enzyme. In many cases, however, for example in the use of microorganisms, while the effects are certainly due to secreted enzymes, they have not been characterized or listed.

ENZYMES

Biotechnology is essentially about enzymes and the way they can be used in food and drink manufacture. Primitive food processing made use of enzymes by using the whole organism. Obvious examples are the use of yeasts in alcoholic fermentations, where a series of enzymes come into play to turn a variety of carbohydrates—mainly sucrose, glucose and fructose—into ethanol. The enzymes are conveniently packaged inside the yeast cells, but the processes are certainly enzyme mediated.

Possibly the next step towards modern conditions is the use of fermentations where the enzymes involved are secreted. The *Aspergillus*-mediated conversions of soy beans to produce a whole range of fermented foods depend on enzymes acting outside the cell, though several acting in concert are involved. There appears to be a subdivision of this case where the enzymes remain attached to the cell walls of the organism, but these are more likely to be important in large-scale fermenters for chemical feedstock production rather than in processing the typical highly viscous food product. Glucose isomerase (Chapter 2) is a notable exception.

Next is the use of specific enzymes, apart from the producing organism. The oldest example is probably rennet (also called 'chymosin'), a protease obtained from the calf stomach and which is used in cheese manufacture. Thus there is certainly nothing new about the idea of enzymes or even isolated specific enzymes in food manufacture.

There are, however, a number of attractive processes which could be carried out by enzymes if they were available. Starch to maltose, glucose to fructose, modification of some gums to improve their properties, modification of lipids and proteins in specific ways—all offer opportunities of improving processes and products. Some of these processes are now in use but they were all dependent on the availability of enzymes in sufficient quantity. The techniques of biotechnology now make virtually any enzyme potentially available in whatever quantity is needed. This is done by transferring the appropriate gene into a host organism which is then further induced to produce the relevant enzyme, or other protein, in as large a yield as possible. The ability to do this offers another possibility which is at least as important. The presence of an enzyme in the host organism can be used

The Basic Operations of Biotechnology

to modify its metabolism, and if it is for example a leguminous seed or an oil seed, it might be used to modify the composition of very important raw material sources for the food industry. Although initially the genes that are likely to be used will be those already in existence, in some organism somewhere, the techniques are not restricted to these and it is quite possible to use modified or even completely synthetic genetic material in the host organism.

Thus providing we can specify an enzyme to do the job, we are not restricted to those already in existence. At the moment, insufficient knowledge prevents this except in one or two heavily studied cases, such as the proteases. In another example, enzymes with the ability to remain active at high temperatures, the evidence so far collected suggests that such specification will be very difficult indeed. Attempts to modify the metabolism of the host organism are not likely to be simple. There are very few organisms where the biochemistry is known in sufficient detail to be able to predict the effect of inserting an extra enzyme. It should be noted that this is not replacement, which would also require the selective deletion of a gene. This is actually quite difficult to do, although uncontrolled random deletions have been used in the production of bacterial mutants for a long time. For higher organisms, one has to rely on chance discovery of strains lacking genes of interest. For example, quite recently soy beans lacking lipoxygenase have been found and these might then be a good choice as a host organism.

Time and Cost

A characteristic of the individual steps of biotechnology is that they are more than ordinarily sequential, i.e. the first step must be completed before the second can be started, and so on. They also require a collection of experimental skills and specialization that is unlikely to be found in a single team. Table 1.1 gives a

Table 1.1 Steps in biotechnology

Stage	Time	Teams
1. Basic biochemistry, identification of enzyme	2–3 years	1 × A
2. Isolation of enzyme, molecular weight	1 year	1 × A
3. Partial sequencing, nucleotide probe synthesis, isolation of mRNA, DNA synthesis via reverse transcriptase, DNA sequencing	1.5–2.0 years	1 × B
4. Incorporation into vector, insertion into host organism, propagation of host,	2–3 years	2 × C
isolation of enzyme or processing of modified source		1 × A

*A, B and C indicate three teams with different expertise required.

summary of the individual steps in a strategy for identifying a suitable enzyme and transferring its gene to a host organism. Making the assmption that a well-found laboratory meeting all the regulatory requirements for recombinant DNA work already exists, and that experienced and skilled research workers are available, it requires between six and nine years to complete the operation. It can be shortened to three to six years if the enzyme in question is well known and fully characterized, while if a flowering plant is to be the host organism, Stage 4 may be considerably longer since only very limited success in transferring genetic material into plants has so far been achieved.

The effort required is given in 'team' units (see Table 1.1). This is the basic functional unit of research and consists of a post-doctoral or experienced team leader, a couple of graduate assistants and a technician. The total annual cost of such a team, including support staff and accommodation is at present about £100 000 ($160 000). Four teams will be required. Thus, the total cost is £2–4 000 000 in revenue costs alone. However, because of the sequential nature of the work, it is hopelessly uneconomical to set up such a unit to undertake only one operation, and once the pipeline is full it could do four simultaneously. This considerably reduces the cost, but it is unlikely to ever average < £1 000 000 per gene. To put this in a commercial perspective, this is the profit to be made from the sale of 5 000 000 fish fingers, which if laid end-to-end would stretch 500 km!

Most commercial organizations will happily spend a million pounds if they can see an annual return, either in profits or savings, of around £200 000. Thus, if they could save this amount in, say, ingredient costs by the judicious use of an enzyme-based process, then there is a case for spending up to a million pounds in obtaining that enzyme. But some caution is needed. The estimates above are for the cost of obtaining an organism containing the appropriate gene and producing its product. A good deal more than this would be needed to build an actual production plant on the scale needed. It is usually held that the steps to production plant cost about ten times as much as the initial research phase, but this is a mixture of revenue and capital expenditure, albeit involving many miles of fish fingers to raise! Quite elaborate costing schemes, including discounted cash-flow calculations which essentially include the cost of money, lead uniformly to the conclusion that food-stuffs or ingredients having an annual turnover of < £5 000 000 ($10 000 000) are unlikely to justify a biotechnology exercise and on financial grounds alone, £10 000 000 would be more comfortable. This rules out a number of minor food ingredients but fortunately leaves plenty of scope in what is, in one way or another, the world's largest industry. There will be some examples of small markets in Chapters 5 and 6. In our discussion of general principles and the potential and practical problems of biotechnology, we will now consider the vast tonnages and financial flows of some sources of lipids for human consumption.

Modification of Lipid Sources

Biotechnology is potentially capable of having widespread effects on such an

The Basic Operations of Biotechnology

Table 1.2 Production of lipid sources (million tonnes)

Source	Area	1979–81 (mean)	1984	1985	1986
Rape seed	World	11 131	16 592	19 070	19 641
	North America	2584	3430	3509	3888
	Europe	3170	5869	6271	6559
	UK	274	925	895	950
Soy	World	86 018	90 233	100 575	95 521
	North America	56 095	52 285	58 776	56 230
	Europe	624	916	919	1608
Sunflower	World	14 397	16 465	19 119	20 804
	North America	2536	1799	1532	1287
	Europe	3087	4484	4779	5998
Groundnut	World	18 352	20 145	21 307	21 512
	North America	1738	2235	2053	1841
	Europe	24	21	21	20
Coconut	World	3690	4032	3949	3882
Palm oil	World	5024	6932	7629	8226
	USA	44	59	73	71
Kernel	World	1751	2409	2619	2752

enormous and diverse field, but financial considerations, although important, are not the only ones involved.

The total world production and the tonnages for some geographical areas are given in Table 1.2 for a selection of the most important lipid sources for human food. They are all also important in international trade. Soy beans were, after wheat, the largest internationally traded agricultural commodity, though palm oil now rivals it. Clearly, there are sufficient cash flows to make financing the research quite feasible, while the tonnages are such that only marginal improvements would give a sufficient return.

The reason for the extensive trade is that the main consumers happen to live in temperate zones where the major sources cannot be grown.

As can be seen from Table 1.2, palm-oil production has increased sharply in the last few years and the introduction of élite clonal palms together with the spread of oil palm plantations indicates a continuing trend. The introduction of erucic acid-free rape seed has also led to a substantial increase in production for human food, particularly in the UK. It is interesting to note that this major change was based on classical plant-breeding methods, which have by no means exhausted their potential.

The growth of the world's population, the formation of the European Economic Community with a controlled agriculture, and many other factors, all point to a

Table 1.3 Fatty acids found in major lipid sources

Source	Lipid in seed (%)	C_6	C_8	C_{10}	C_{12}	C_{14}	C_{16}	C_{18}	$C_{18:1}$	$C_{18:2}$	$C_{18:3}$
Rape	42	–	–	–	–	–	4	2	64	–	9
Soy	21	–	–	–	–	1	11	4	22	53	8
Sunflower	40	–	–	–	–	1	6	3	23	64	1
Groundnut	45	–	–	–	–	–	8	3	56	26	–
Coconut	66	1	7	7	48	17	9	2	6	3	–
Palm oil											
mesocarp	55	–	–	–	1	3	43	4	40	8	1
kernel	47	–	4	5	47	16	9	2	18	1	–

shift in the raw material sources for manufactures of lipid-based products. Unfortunately, we cannot simply replace soy-bean oils with rape or palm oil with sunflower. The reason for this is given in Table 1.3, where the fatty acids found in the storage lipids of a few typical oil seeds are shown. (Extensive collections of such data are available.) They show quite marked differences in two respects—the chain length of the fatty acids and the degree of unsaturation.

It is far from obvious why different fruits and seeds should contain such a wide variety of triglycerides, and there are no easy explanations of evolutionary advantage. Tropical fruits such as coconut contain saturated triglycerides melting at about 25 °C while the temperate oil seeds tend to be longer chain length but unsaturated, so that they melt at similar temperatures. However, palm kernel and mesocarp contain quite different triglycerides within the same fruit.

It is likely that storage triglycerides evolved as over-produced cytoplasmic fats. These are found mainly in membranes where their precise composition is known to affect the fluidity and functional efficiency of the different kinds of membrane found in the cell. Phospholipids are also a potential ancestral source of triglycerides since they are on the synthetic pathway. Thus one might anticipate that there will be a number of points in the detailed mechanism for triglyceride synthesis where alteration could lead to over-production of C_{10}, C_{12}, C_{16} and C_{18} fatty acids, though apparently not C_8, or less, since there are no examples of these as the principal storage lipids. When 12 different clonal oil-palm lines were analysed, it was found that they fell into two groups with slightly different fatty acid patterns. This indicates that a single gene difference can bring about such shifts, and that there is actually a population polymorphism. No studies on this aspect have been done on common lipid sources as yet.

The presence of unsaturated fatty acids is easier to explain since they only occur where specific desaturase enzymes are found. Also, chain lengths $>C_{16}$ are produced by a distinct chain-elongating enzyme and do not occur when it is absent. However, the question of why most plant triglycerides are based on C_{16} or C_{18} acids, but why some are predominantly C_{12}, remains without a precise answer. It is nevertheless an important question because food manufacturers and the chemical industry have developed products such as margarines, and soaps with quite specific fatty acid chain length requirements. There are, in addition, requirements for unsaturated fatty acids, since many people now believe that a higher proportion of unsaturated fat in the daily lipid intake is desirable. There are also requirements for specific triglycerides for chocolate manufacture (described in more detail in Chapter 5) and others would doubtless emerge if the raw materials were available. Thus the long-term need is for oil seeds which can be grown if necessary in temperate climes, and which are able to produce storage triglycerides rivalling in variety those at present obtainable from all parts of the world.

In biotechnology terms, can we, by moving enzymes from one species to another, control the kind of fatty acids found in the storage triglycerides? We have now reached the first stage in the scheme set out in Table 1.1 and must review the biochemistry and see if a suitable enzyme or enzymes can be identified.

STAGE 1. BIOCHEMISTRY

Control of Chain Length in Fatty Acid Synthesis
Fatty acid synthesis occurs in a great variety of tissues and requires acetyl and malonyl CoA as substrates. The initial step is the transfer of these groups from the thiol group of CoA to the thiol group of the pantotheinyl residue of an acyl carrier protein (ACP). In animals and some fungi and bacteria this is incorporated into the peptide chain of the fatty acid synthetase complex, while in plants and most bacteria ACP is not covalently linked and can be isolated as a separate small protein. The sequences of some ACPs are given in Fig. 1.1. The structure of the acylation site is shown in Fig. 1.2 and both ACP and CoA contain the B vitamin, pantothenic acid. The transfer is mediated by transferase enzymes, and usually two different enzymes are involved for the acetyl and malonyl transferase reactions.

As might be expected such a transfer reaction involves acylation of the transferase enzymes. The active centre sequences for the yeast enzymes are as follows. For the malonyl transferase,

Ala–Gly–His–Ser–Leu–Gly–Glu–Tyr–Ala–
$\qquad\quad\;\;|$
$\qquad\quad\;\;$O
$\qquad\quad\;\;|$
$\qquad\quad\;\;$CO·CH$_2$·CO·OH

and for the acetyl transferase,

Ser–Leu–Gly–Leu–Thr–Ala
|
O
|
CO.CH$_3$

and while they have similarities, they are clearly different. In yeast FAS (fatty acid synthetase), which produces CoA derivatives of fatty acids, unlike mammalian FAS which produces free fatty acids, it is known that the malonyl transferase is the same as the palmitoyl transferase which moves the products of synthesis back to CoA. Plants probably produce ACP fatty acids and in the situation we are primarily interested in, of storage triglyceride synthesis, they are channelled straight into the triglyceride synthesis pathway.

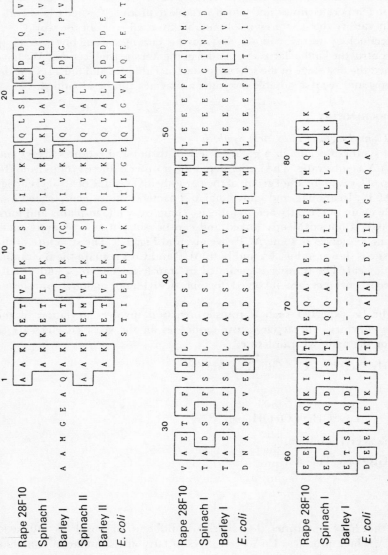

Fig. 1.1. Sequences of plant acyl carrier proteins. Only partial sequences are available for spinach and barley II. Sequences are aligned on serine 39, the site of pantotheine attachment. (From R. Safford et al. (1988). Eur. J. Biochem. **174**, pp. 287–295.)

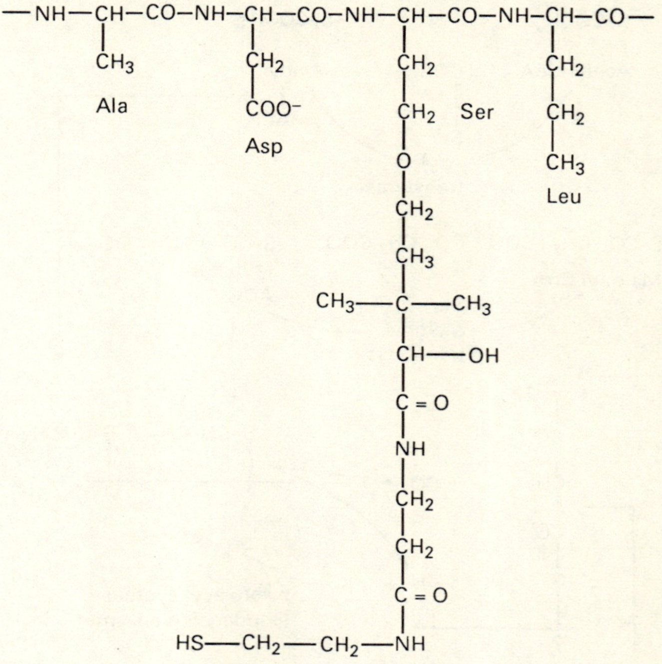

The Basic Operations of Biotechnology

```
—NH—CH—CO—NH—CH—CO—NH—CH—CO—NH—CH—CO—
     |           |           |           |
     CH3         CH2         CH2         CH2
     Ala         |           |           |
                 COO⁻        CH2   Ser   CH2
                 Asp         |           |
                             O           CH3
                             |           Leu
                             CH2
                             |
                             CH3
                      CH3—C—CH3
                             |
                             CH—OH
                             |
                             C=O
                             |
                             NH
                             |
                             CH2
                             |
                             CH2
                             |
                             C=O
                             |
            HS—CH2—CH2—NH
```

Fig. 1.2 The acylation site of acyl carrier proteins.

The next step involves the condensing enzyme, which makes a ketoacyl fatty acid and carbon dioxide by the addition of the malonyl residue at the carboxyl end of the chain. Figure 1.3 shows the reaction with acetate to give the ketobutyryl derivative. The scheme shown in Fig. 1.3 suggests that the residues are moved to thiol groups on the condensing enzyme. While one thiol is certainly involved, in general the evidence for two closely adjacent ones is as yet only for yeast, and one of these may be an ACP. Transfer back to ACP, like the loading of the enzyme does not involve a separate transferase activity. The remaining steps of reduction, dehydration and a further reduction almost certainly occur with the acyl group remaining attached to the pantotheine chain. This has led to the charming idea that this acts as a sort of swinging arm round a ring of activities, which is supported by no concrete evidence, but is nevertheless widely accepted. Three distinct activities are involved, though in some plants there are known to be two distinct reductases at each step, one using NADPH, and the other NADH giving the impression that the plant fatty acid synthetase system, which is located in the plastids, is made up of the remnants of at least two distinct synthetases. As is illustrated in Fig. 1.3, the first turn of the cycle produces butyryl ACP. Figure 1.4 shows the next step, where further incorporation of malonate produces the hexanoic acid, and so on. At each turn the acyl ACP has an option—it can either

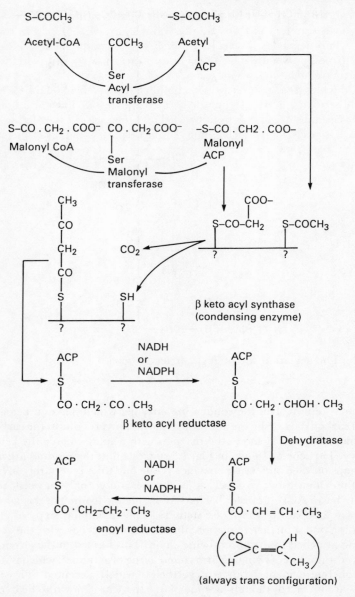

Fig. 1.3 Enzymes involved in the synthesis of fatty acids.

The Basic Operations of Biotechnology

go to the condensing enzyme for elongation, or it can transfer to CoA, to water, or in plants to a transfer ACP, or simply wander off away from the rest of the enzymes. (This depends on whether plant FAS is a complex enzyme or not, an unresolved question discussed below.) The option exists up to C_{16}, palmitic acid, at which further elongation cannot occur.

This is the first clear point at which chain length specificity can be seen and it presumably resides in the structure of the condensing enzyme. This is not absolutely certain since, as has been pointed out, the malonyl transferase enzyme, at least in yeast might have such a high binding capacity that it simply pre-empts all the available palmitoyl ACP. In other species such as rabbit, there is a distinct thioesterase activity responsible for removing the synthesized chain from the

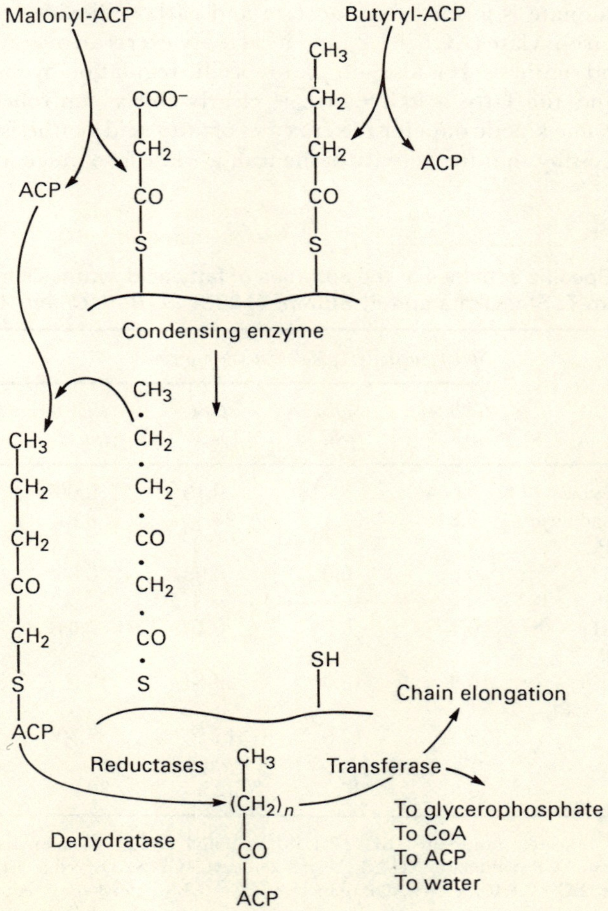

Fig. 1.4 The chain lengthening reactions in fatty acid synthesis.

pantothenate residue of the ACP moiety, and such an enzyme may be found in all fatty acid synthetases where free fatty acids are the product. There could be a chain length specificity in this.

In vitro experiments do not result in the synthesis of a single fatty acid such as palmitate. Typically C_8, C_{10}, C_{12} and C_{14} acids are also made in significant amounts, and one must suppose that at the decision point, shown in Fig. 1.4, the probability of elongation as opposed to release is determined by the affinity of the condensing enzyme as opposed to the thioesterase. The balance of these changes with chain length until at C_{16} it shifts entirely to the thioesterase. There is also a kinetic component, since varying the ratio of malonate to acetate changes the relative proportions of the chain lengths made. Obviously if a growing chain arrives at the condensing enzyme, and fails to find a malonate waiting for it, this increases the chance that it will be released rather than elongated.

In vivo, malonate is formed from acetate and carbon dioxide by the enzyme acetyl CoA carboxylase (EC 6.4.1.2) and its activity exerts an overall influence on total fatty acid synthesis. It is also subject to specific regulation by palmitic acid in mammals, and the fatty acid balance is clearly finely controlled. Table 1.4 summarizes some kinetic data for the enzymes of fatty acid synthesis for five plant systems, suggesting that the activity of the transacylase is so much lower than the

Table 1.4 Specific activities of the enzymes of fatty acid synthesis in some plants. (Adapted from T. Shimakata and P. Stumpf (1983). *J. Biol. Chem.* **258**, p. 3592.)

Enzyme*	Specific activity (nmol/min/mg protein)				
	C. lutea	Safflower seeds	Rape seeds	Pea leaves	Spinach leaves
Acetyl transacylase	0.064	0.018	0.19	0.005	0.009
Malonyl transacylase	9.8	24.1	24.4	8.81	12.3
Ketoacyl-ACP synthetase I†	0.6	0.05	0.087	0.05	0.02
Ketoacyl-ACP synthetase II†	0.10	0.04	0.10	0.04	0.15
Ketoacyl-ACP reductase	16.8	18.9	36.6	13.2	19.2
Hydroxyacyl-ACP dehydrase	8.3	12.8	15.2	7.30	6.72
Enoyl-ACP reductase	36.9	32.7	42.4	30.9	32.4

*Acetyl (ACP) S-acetyl transferase (EC 2.3.1.38); malonyl (ACP) S-malonyl transferase (EC 2.3.1.39); ketoacyl-ACP synthetase (EG 2.3.1.41); 3-ketoacyl-ACP-NADP reductase (EC 1.1.1.100); enoyl dehydrase (EC 4.2.1.17); enoyl-ACP reductase (EC 1.3.1.9); enoyl-ACP reductase (NADPH) (EC 1.3.1.10).
†Synthetase I is the palmitate limited condensing enzyme and synthetase II the palmitate stearate elongase.

others that it is likely to be the rate-limiting step. Adding extra acetyl transacylase caused an increase in the amount of fatty acid made, but reduced the average chain length.

Chain Elongation. Although both plant and animal FAS make palmitate, a distinct synthetase which uses CoA or ACP palmitate and malonate as substrates is very widespread and makes stearic acid C_{18}. Other elongases also occur and are able to produce C_{20} and C_{22} acids with the aid of distinct enzymes. It is the elimination of a C_{20} elongase by classical breeding methods that has made erucic acid (C_{22})-free rape seed available (see Table 1.2). It is not clear whether these systems require only a different condensing enzyme with the rest of the complex being the same as for palmitate synthesis. In animals, since all the activities are on the same peptide chain, it must have a complete set, but it is not known whether they differ significantly. It is claimed that elongases use only NADPH which would differentiate them.

Clearly, these enzymes have a marked chain length specificity which almost certainly resides in the condensing enzyme, and possibly the rest of the system as well. So far, no plant condensing enzyme has been isolated, still less sequenced and it will be some years before the specificity can be understood in terms of the enzyme structure. There is therefore little chance in the foreseeable future of influencing fatty acid chain length by careful alteration of the ketoacyl synthetase, though this remains, in principle, an attractive route.

There are some examples in plants worth investigating. The chain-length limiting mechanism of palm kernel is unknown, and might yield an interesting condensing enzyme. Some work has been done on coconut endosperm (*Cocos nucifera*) which has about 50% C_{12} and C_{14} fatty acids. Oddly, cell-free extracts made mainly C_{16}, palmitate, and stearate C_{18}, though tissue slices made mainly C_{12} and C_{14}. Specificity of acyl transferases was ruled out as a mechanism and the matter remains unresolved, though kinetic control combined with compartmentalization of enzymes seems the most likely explanation. *Cuphea*, with mostly C_{10} fatty acids is another possible source of regulating enzymes, and there remains much work to be done before biotechnology can be brought into action.

Work with plant fatty acid synthetases is difficult because of the requirement for ACP. Until recently, the only ACP that had been characterized was the *Escherichia coli* protein: it remains the only one available and while it certainly functions in plant systems, there must always be doubt about detailed comparative data obtained with it on, for example, chain-length specificity. It cannot be purchased unlike a wide variety of CoA derivatives and must be isolated from several kilograms of *E. coli*—a difficult preparation. It is now known that plants contain at least two different ACPs, probably with slightly different functions. Rape seed codes for at least five different ACPs in its nuclear DNA, though it is not yet known how much of this is a population polymorphism.

It may be that a genetic engineering exercise to produce a good supply of ACP will be necessary before much more progress can be made. ACP itself is a candidate for a chain length regulator. *In vitro* addition of ACP to fatty acid synthetase reduces the average chain length made. At least one group is now attempting to insert spinach ACP genes into rape seed in an attempt to modify the lipid composition. The gene has been cloned in *E. coli*.

The reductases in the cycle in plants and in *E. coli* can both use CoA derivatives as substrates, which makes them much easier to isolate and work with than condensing enzymes. This is why these have been chosen for isolation and sequencing with a view to isolating the gene and the rest of the fatty acid synthetase complex (if there is one, see below). The target is the FAS of the seed, activated during maturation. This avoids the need for ACP but does not eliminate three other particular experimental difficulties: cellulose cell walls mean not only that tissues are difficult to extract, but the protein content is lower than in animal tissues. Secondly, maturing seeds are apt to be small, and only obtainable in significant amounts once a year. Thirdly, plant tissues have notably active proteases, for which no good inhibitors exist. FAS itself is inhibited by the same reagents that inhibit thiol proteases such as papain.

This goes some way to explain why relatively little structural information on plant FAS is available. More important it reinforces the conclusion reached above that it will be some years before much progress is made.

Structure of Fatty Acid Synthetase
It is evident that while the substrate changes leading to fatty acid synthesis are well established and universal, the enzymes that bring them about are little known and show a wide variation.

Some details of known structures are given in Table 1.5. The main question is whether the activities are on one, two or more peptide chains, and how many complete sets of activities make up the functional molecule. As can be seen, there is some variety, and at least three types can be distinguished. It is sometimes said that prokaryotes and plants have a synthase made up of six or seven separate enzymes, while eukaryotes have complex enzymes, but this is an over-simplification. First, complex enzymes can exist without the various activities being on the same peptide chain. In fact subunits are usually held together by non-covalent interactions, and the mammalian FAS, with six activities and the ACP moiety all incorporated into one peptide chain is distinctly unusual amongst enzymes. (It actually raises some fundamental questions about how such an enzyme can fold correctly. One supposes that the domains form as the growing chain emerges from the ribosome. If this is not what happens, and some external agency, such as the right kind of membrane, or even another protein such as rearrangease is needed, then moving the gene to another organism for expression is not likely to succeed.)

An active site seems to require a molecular weight of about 20–30 000 to sustain it. A great many enzymes fall within this range so the mammalian FAS represents a very efficient use of peptide chain with seven active sites in 240 000 molecular weight.

Complex enzymes are kinetically more efficient, since the active centres are so close together. The effect is to optimize the substrate concentration at all times. On the other hand, molecules of this size are more susceptible to protease attack: they appear to consist of domains with relatively accessible loops linking them together. Proteases can be used to partly digest the complex, and the different activities can

Table 1.5 Structures of some fatty acid synthetases

Organism	Peptides	MW*	Sets	Product	Chain lengths	Comments
Yeast	2 (α and β)	1 200 000	6	Acyl CoA fatty acids?	14, 16	Contains FMN
Chicken liver	1	500 000	2	Acyl CoA	14, 16	and many other avian and mammalian livers
E. coli	6+ACP	Complex? 250 000	1	Acyl ACP fatty acids?	14, 16 18–24	
Cyanobacter						
Mycobacter smegmatis	1	1 700 000	2	Acyl ACP	14, 16	Contains FMN, 10 and 12 in some tissues
Plants, e.g. safflower	6+ACP	Complex? 250 000	1			
Aspergillus fumigatus	2 or 6?	1 500 000				
Mammary glands (rat)	2+2	600 000				
Uropygial gland (duck)	2+2	600 000	2	Fatty acids	8, 10, 12	Medium-chain hydrolase present
Pea aphid	2 ?	?				
Ruminant mammary gland	1	500 000		Fatty acids	8, 10, 12	No distinct hydrolase
				Acyl CoA	8, 10, 12	1 Pantotheine in two chains
Ceratitis capitata (fly)	2	500 000		Fatty acids	16, 18	
Drosophila melanogaster	?	?		Fatty acids	12, 14 or 16, 18	Ionic strength-dependent

*Molecular weight of functional enzyme.

be separated. Extensive nicking can occur without affecting the overall activity, and this was not fully appreciated in earlier work on these enzymes, where activity was the sole criterion of the integrity of the enzyme. *In vivo* there is some evidence to suggest that longer chains have a shorter turnover time.

In the case of FAS, the advantages of complex enzymes are compounded by the fact that the functional enzyme often contains two or more complete sets of enzymes in close proximity. In yeast, which has six sets, the growing chain can move from one set to another, depending on which one happens to be free and provided with a malonyl residue, and this also occurs in the mammalian binary enzyme.

None of this solves the question of whether the *E. coli* and the plant FAS exist as complex enzymes. Careful work has made it unlikely that the fact that the different activities can be isolated as clearly distinct peptide chains is the result of proteolysis. There is no hint at all of complex formation, or of the association of FAS activity with large molecules of the order of 500 000 molecular weight in tissue extracts.

Plant metabolism proceeds a good deal more slowly than animal, and the rate of deposition of lipid in maturing seeds is such that a quite inefficient enzyme would be adequate. It is, for example, much slower than the rate of lipid synthesis in yeast. If ACP is seen as a kind of detachable subunit, able to migrate from enzyme to enzyme, and bearing in mind localization in the plastids, a dispersed FAS might be quite effective. There remains a possibility of partial complex formation. The four core enzymes—that is the condensing enzyme, the two reductases and the dehydratase—together with a specialized ACP form, is the most probable functional arrangement. Evidence is completely lacking as yet.

From the point of view of biotechnology, the important question is how many chains, because this should indicate the number of genes involved. The mammalian FAS is all specified by one mRNA. Yeast requires two genes for its two peptides, and they are actually carried on different chromosomes. Attempts are now being made, via the relatively accessible reductases of the plant system, to locate their genes, but whether the remainder of the enzyme complex will be found in close proximity remains to be seen. The sequence of mammalian FAS by DNA sequencing is likely to appear soon.

Desaturases

The unsaturated fatty acids mentioned in Table 1.3 are all produced by the action of desaturase enzymes (EC 1.14.99.6) on the saturated fatty acids. Oleic acid ($9C_{18:1}$), linoleic ($9,12C_{18:2}$) and linolenic acid ($9,12,15C_{18:3}$) are made from palmitic by a specific elongase and a set of desaturases with positional specificity. The double bond is always *cis*. In animals the substrate is always the CoA derivative. In plants it is most likely to be the acyl-ACP, though they can use CoA derivatives *in vitro*. Certain phospholipids can also serve as substrates in maturing seeds.

The desaturases show marked chain length specificity for C_{16} and longer acids, and marked positional specificity for dehydrogenation. They are membrane-associated and require at least some membrane fragments to be present for activity. The enzyme contains, apart from an iron-based dehydrogenase subunit, cytochrome b_5 and a cytochrome oxidase. The membrane-associated nature of the enzyme in plants has made its characterization very difficult, and while it may be possible to approach it through the known structure of cytochrome b_5, genetic manipulation will not be easy. It does not look a good candidate for chain length control, though if in the future the demand for unsaturated fatty acids should increase, it might be possible to transfer the gene. The animal desaturase has been isolated and cloned.

Triacylglycerol Synthesis
Triglyceride synthesis in the maturing seed follows the same general pathway as in the rest of the plant, and in animals. A typical time course for lipid deposition, in this case in sunflower, is shown in Fig. 1.5. There is a major surge of FAS activity which coincides with the appearance of lipid and the enzyme is synthesized, presumably under genetic control, early in the maturation process. In the example shown, FAS activity falls away on maturation but in other species may still be present on final dehydration of the seed.

As suggested above, one reason for locating the gene for any part of FAS is to exploit its genetic control mechanism and target inserted genes to the maturing seed by using it. This has recently been successfully done for pea legumin in tobacco, where specific production in the seed was obtained.

The biochemical pathway to triacylglycerides starts with the acylation of glycerol-3-phosphate to produce a lysophosphatidic acid. This is then further acylated and dephosphorylated to give mono-, di- and tri-acylglycerols, while the phosphatidic acid serves as a pathway to the phospholipids. *In vitro*, the acyl CoA can function in this system, but it seems more likely that *in vivo* ACP derivatives are

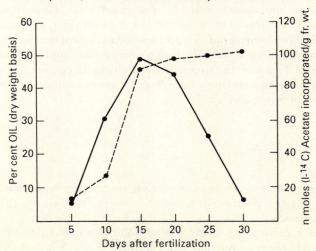

Fig. 1.5 FAS activity and lipid deposition in maturing sunflower seeds. - - - - - - -, percent lipid; ————, acetate incorporation rate. (From P. Monza, S. Munshi and P. Sukhija (1983). *Plant. Sci. Lett.* **31**, pp. 311–321.)

the substrates. There is no evidence of chain length specificity and little is known of the enzymes involved (EC 2.3.1.40).

Conclusions
This brief review of what is known of the synthesis of triacylglycerols in maturing seed does not present a very encouraging picture from the point of view of biotechnology manipulations. Referring back to Table 1.1, Stage 1 would seem to need rather more than two to three years for the elucidation of the basic biochemical pathways, and in particular while there are some indications, it is not possible to point to an enzyme, or even a class of enzymes, as the key to chain length control. The considerable amount of work needed to isolate and characterize the condensing enzyme from, for example *Cuphea*, even if it is within present capabilities, is by no means certain to yield the required insight. Biotechnology to reiterate, is about enzymes, and the most detailed knowledge of substrate pathways without even more detailed knowledge of the enzymes is insufficient.

Thus while information about plant FAS is beginning to accumulate, it is going to be some years yet before the basis of chain length specificity is understood in species like palm and coconut and there is insufficient basic information available to make a biotechnology exercise possible based on plant material alone. Although in the long run the modification of condensing enzymes is probably the optimum way of modifying fatty acid chain lengths in oil seeds, we are not restricted in searching for a suitable enzyme to plant sources. Table 1.4 includes a reference to a specialized chain-limiting mechanism which has been found in some mammals and in the preen glands of ducks. This involves a specific short-chain hydrolase which can force the FAS to terminate at C_{10}–C_{12} rather than its usual C_{16}. Could this enzyme interact with other FAS to give the same result? Experiments *in vitro* show that it does, and these results will be discussed later. The enzyme which we will call MCH for 'medium chain hydrolase' (FAS acyl thioester hydrolase, EC 3.1.2.2) had been partly purified and seemed to present no special problems in obtaining it and ultimately its gene. It also offered what appeared to be the most feasible chance of actually obtaining some chain length control. It turned out to have lots of problems, and thus provides an excellent example of the difficulties that can arise in Stage 2 of Table 1.1.

STAGE 2. ENZYME ISOLATION

Scale of Isolation
The most important question to answer before starting on an isolation is to decide how much of the enzyme is needed, and the source, since this will pre-determine the whole approach.

All enzyme isolations are in one way or another a race against time and the quality and yield of preparations depend on the skill with which the various steps in isolation are integrated together, and the speed with which they can be carried out. Unfortunately, in the present state of the art, it is difficult to exactly reproduce preparations and some variation always occurs. This makes it even more desirable to get the initial estimate right. There are practical advantages in making all measurements on a single sample. Our primary purpose in isolating MCH, or any candidate for genetic manipulation, is to do sufficient sequencing for probe construction and validation of DNA-derived sequences, and also to obtain a molecular weight. Some will also be needed for making antibodies and in demonstrating the successful expression of the gene.

An estimate of individual chain weights can be obtained from calibrated SDS gel electrophoresis (see below) and requires about 50 µg protein. A more reliable absolute value for the molecular weight, which is not necessarily the same as that of the individual chain weight, can be obtained from sedimentation- equilibrium or light-scattering methods and requires about 5 mg, which can however be recovered and used for other purposes. If it is necessary to measure the partial specific volume $\bar{v}$ then at least a further 10 mg will be needed. The amino acid composition can usually be used to calculate an adequate value for $\bar{v}$ but not always for conjugated proteins such as glycoproteins.

Amino acid analysis and sequencing techniques have been developed in recent years to handle smaller and smaller amounts of material. It is sadly true that electrophoresis, which is by far the most powerful method for fractionating and isolating proteins—it is relatively easy to fractionate a mixture of 100 proteins into the individual components, and the current record stands at more than 2000—cannot handle more than about 100 µg total protein at these levels of resolution. After some years of fruitless attempts to scale up electrophoresis, there is now a discernible trend towards developing techniques which can work on the amounts of protein actually available from electrophoretic separations.

Thus, standard equipment can now give amino acid compositions on about 10 µg though precise values need 50 µg. In sequencing, the most recent equipment can give extensive data from 1 mg and in favourable cases less. The present generation of equipment which is still in widespread use would require about 20 mg while the full sequence on a reasonably straightforward example would need up to 100 mg for a typical enzyme of 300 residues and 32 000 molecular weight. Strictly, since longer chains need more material, the requirement is for 3 µmoles.

Antibodies are very useful, but different proteins vary greatly in their antigenicity, i.e. the ease with which they elicit antibody formation. Some highly conserved proteins like actin hardly cause antibodies at all, since the protein is so similar in all species. For a globulin of average antigenicity, about 10 mg is needed to immunize a rabbit and test the antisera. There is an approach via monoclonal antibodies which has been used successfully once. This involves immunizing a single mouse, which needs about 20 µg of protein and then using the spleen for monoclonal antibody formation (see Chapter 6). Although more protein is then needed to screen the clones for an antibody, this can be a relatively crude preparation. The drawback is that most monoclonal antibodies elicited do not have the necessary binding strength, and one has to screen several hundred clones before finding a suitable one. Also for many purposes, particularly isolation and identification work, polyclonal antibodies are preferable though they can be approximated by mixing several monoclonals. Finally, small amounts of enzyme will be needed as authentic samples when proving that the host organism is expressing the transferred gene. Looking still further ahead one might one day want to crystallize the protein for structure determination but that is outside our present horizon. We should however allow a few milligrams for contingencies such as further analysis for carbohydrate content and other kinds of post-transcriptional modification. These are covalent modifications to the structure after the chain has been formed under the control of mRNA on the ribosome. The commonest are attachment of carbohydrate residues, removal of part of the chain, modification of the N-terminal group and formation of intra-chain disulphides— 5 mg should suffice. The total comes to about 50 mg as a suitable target for MCH, or for the FAS enoyl reductase of rape seed, both of which are 300–400 residue globular enzymes.

Source

It would be ambitious to plan for anything more than a 10% yield, so we must find tissues containing 500 mg of MCH. In most laboratories, lactating rats are more readily available than mallard ducks, or large weights of aphids, and published data suggests that 500 mg MCH would be found in the mammary glands of about 60 rats, weighing around 1 kg. This brings the isolation just within the range of small-scale laboratory equipment, which is limited by homogenizer and centrifuge capacity of 5–6 litres. In contrast, a similar calculation for the FAS enoyl reductase suggests that at least 10 kg of maturing rape seed would be needed, which would require small pilot-plant scale equipment and a quantity of seed which is barely feasible. For plant sources it is essential to use every possible device to minimize the amount of protein needed.

Extraction and the First Fractionation

The objective of this stage is to get the enzyme in solution and carry out the first fractionation as quickly as possible to eliminate cytoplasmic protease attack. This stage is often done at reduced temperature: this may slow down proteases but is more important in minimizing bacterial and fungal growth. These organisms are also sources of proteases. The pH is probably the most significant variable at this stage. With plant sources it is advisable to measure protease activity as a function of pH. There is often a 'window' at pH 3.5–4.0 where protease activity is minimal and extraction at a suitable pH can greatly improve yields.

It is necessary to break cell walls and often disrupt organelles as well. For animal tissues, simple blade homogenizers are available for a 5 litre scale and there are no special problems. Plant and bacterial cells are very different, and there is no really effective large-scale extraction method for plant tissues. High-speed blade homogenizers, or even grinding with sand, may have to be used in repeated small batches. Bacteria and yeasts can be effectively disrupted by homogenizers of the Manton–Gaulin type. In these, a suspension of cells is forced through a nozzle under very high pressure. For large-scale work, machines used in ice-cream manufacture can be adapted and 5 kg can be smashed in 30 min. All these homogenizers have problems of effective cooling of the suspension.

Centrifugation was chosen by the pioneers of biochemistry for removal of insoluble debris because it is quick compared with other methods. Small flow through centrifuges are best for this stage when 10–20 litres are involved, but are rarely seen in laboratories. The upper limit for most laboratory centrifuges is 5–6 litres and they are not particularly efficient at these volumes.

It may be best to use a filtration step with a filter aid such as celite which can be equally rapid and more efficient for the 5 litre scale extraction. The ratio of extractant to tissue volume determines the efficiency of the recovery of the supernatant and plant tissues which have very bulky residues compared with animals need a higher volume of extractant, and produce relatively dilute extracts.

There can be advantages in performing a cell fractionation, for example to isolate plastids or protein bodies, before extracting the enzyme. Unfortunately, the scale of operation in standard procedures is much too small to be worth considering for the 50 mg yields we need. There are also methods of variable utility using cellulases to break down cell walls, which may help extraction but complicate subsequent fractionation.

Enzyme activity in these primary clarified extracts disappears more or less quickly for two main reasons. The protease attack already mentioned, but also some enzymes, will interact with other proteins to form inactive aggregates. Providing the pH is kept within the range 3.5–9.0 and ionic strength in the physiological range of 0.1, spontaneous structural alteration—'unfolding'—is uncommon at 25 °C. Thiol reagents such as dithiothreitol should not be added unless there is evidence that they do actually preserve the enzyme activity, since they can promote disulphide rearrangement and aggregation (see Chapter 4). Protease inhibitors can be useful. Phenylmethylsulphonylfluoride is often used, but has its dangers. It neither inhibits all proteases nor is it specific for them. It inhibits thioesterases such as MCH. We must therefore remove the unwanted components of the extract as quickly as possible and in addition (since all subsequent operations will require a much smaller volume of extract), we have to concentrate the enzyme of interest.

Further Fractionation

There are three ways of preparing the initial extract for further fractionation. The oldest, and still one of the best, is to precipitate the protein by adding large amounts of a salt, usually ammonium sulphate, or an organic solvent such as ethanol, followed by centrifugation to collect the precipitate. This works very well for volumes up to 1 litre but over this centrifuges are less efficient than filters. The precipitate after collection is dialysed into the appropriate buffer for the next step. This is slow, but the enzyme is usually stable at this stage in the preparation. Some other possible strategies are outlined in Fig. 1.6. They all have advantages and disadvantages, and combine four basic fractionation methods in different ways.

1. *Ion-exchange Chromatography.* The prototype materials are carboxymethyl cellulose, bearing cationic COO^- groups and diethylaminoethyl cellulose bearing basic $CH_2CH_2N^+(CH_3)_2$ groups, but also some COO^- groups. Others have been developed based on different materials including cross-linked dextrans and silicates, but apart from flow characteristics when used in columns, they all have similar fractionation capabilities.

Proteins adhere to the column material when their net charge is the opposite of

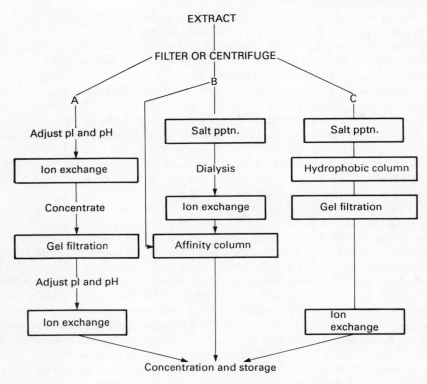

Fig. 1.6 Strategies for enzyme isolation. Numerous combinations and permutations of the four basic fractionation methods are possible. The three shown are commonly used.

the column, though the strength of interaction depends on the ionic strength. Experience suggests that it is best to use these columns at a pH fairly near the isoelectric point (pI) of the protein. For MCH, for example, with a pI = 5, it would be appropriate to use DEAE cellulose at about pH 6. Proteins above their pI have a net negative charge. The pI is a quantity it is very helpful to know, though it is rarely available in advance of fractionation attempts. Elution is then performed by raising the ionic strength. A typical elution profile, for MCH extracts is shown in Fig. 1.7 where a linear gradient was used. The basic experimental arrangements of a column, with effluent monitoring by UV absorbance and a fraction collector, have changed very little since the method was introduced in 1950 though numerous automated commercial systems are now made and large-scale (i.e. 50 litre) equipment, while not common, is available. Figure 1.8 shows the basic arrangement which is also adopted for large-scale industrial plant.

Elution by varying the pH is less easily controlled and reproducible. Because protein molecules are large and carry many charged groups, they can bind to the column in a large number of energetically nearly equivalent ways which

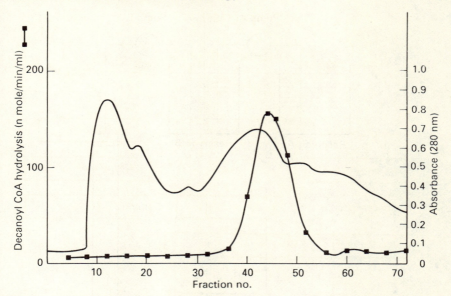

Fig. 1.7 DEAE cellulose chromatography of a crude extract containing MCH. The column was packed in 0.01 M sodium phosphate pH 7.0 and elution was with a linear gradient rising to 0.4 M NaCl in the same buffer. There are at least 500 proteins present and it is not surprising that the activity peak does not coincide with any of the protein peaks as determined from UV absorbance.

inevitably limits the resolution obtainable. The advantages of the method are that it can accept fairly large volumes of applied protein solution, providing the pH and ionic strength are correct, and the total protein capacity is high. A disadvantage is that fractions emerge in rather large volumes and have to be concentrated for the next step.

2. Gel Filtration. This technique relies for its discrimination on the way in which the size of a protein molecule limits its ability to diffuse into pores of the same order of magnitude as the molecular diameter. Column materials with a variety of pore sizes are available and if a mixture of proteins is applied to the top of such a column and then slowly pumped through it, the small molecules are able to diffuse into the pores faster than the large ones. Effectively, the column has a larger volume for small penetrating molecules than for large excluded ones. Large molecules therefore emerge first.

There is a further important consequence. Excluded molecules emerge from the column in the buffer it was packed in, and not the one they were applied in. Thus an advantage of this method is that it can be used to change the solvent, which is equivalent to a very rapid dialysis.

The main disadvantage of the method is that it can only accept small volumes,

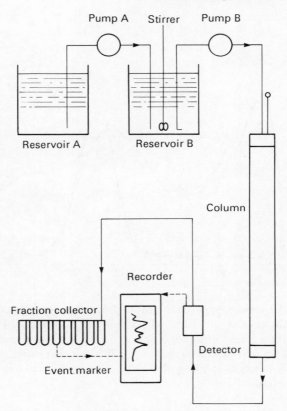

Fig. 1.8 Typical arrangement for column chromatography of proteins. By adjusting the rate of pump A and B and the composition of the two reservoirs a variety of linear gradients can be produced. More complex arrangements are also used. After the column a detector measures some property of the effluent: this is usually the UV absorbance, at wavelengths between 210 and 290 nm depending on the concentrations. Refractive index and conductivity have also been used. The fraction collector is linked to the recorder so that the record can be correlated with the fraction. Samples are applied to the column through the pump, and pressurized systems (HPLC) are now becoming widely used.

and limited amounts of protein. Though it can sometimes be used in combination with salting out (Fig. 1.6C) it is best used as a final step to control the solvent for storage purposes. The resolving power is not high, and is actually better than might be expected since molecular diameters are related to the cube root of the molecular weight and only double over the range 10 000–100 000 usually encountered.

The Basic Operations of Biotechnology 27

3. *Hydrophobic Chromatography.* Introduced recently for proteins, this is an ingenious adaptation to columns of the basic principles of salting out. The columns are cross-linked dextrans, bearing octyl or phenyl groups. Protein is applied in ammonium sulphate or other high ionic-strength medium and eluted by reducing the salt concentration. It may be necessary to use ethylene glycol to elute, and some enzymes cannot be eluted at all under mild conditions. Because they bind from strong salt solutions they are particularly well suited to follow an initial ammonium sulphate precipitation step. A relatively uncommon adsorption material is calcium phosphate gel, which can be used for protein fractionations. Its main limitation is small capacity for protein. As it happens, it was used in an isolation of MCH.

4. *Specific Affinity Columns.* There are two kinds of these. The first which uses antibodies is of relatively little interest here since in order to make antibodies it is necessary to isolate the enzyme first. However, it may sometimes be possible to use antibodies already raised against a related enzyme and which cross-reacts. This is a common property of polyclonal antibodies which are not usually very species-specific. Thus, antibodies against *E. coli* ACP will cross-react with rape-seed ACP.

Techniques exist for coupling antibodies to matrices such as dextrans or silicates so that they can be used in column form. The major drawback is that it is difficult to elute the enzyme in an active form, and there is also a risk of inactivating the antibody. In our examples, no antibodies were available.

A much more useful approach uses column-linked groups that are similar in size and shape to either the substrate or coenzyme of the enzyme in question. It has been found that enzymes will interact strongly with such groups and they can easily be eluted by adding the substrate. There are many examples, one of the best known being the use of triazine dyes, such as cibacron blue, to affinity bind NADH-dependent reductases. This was used successfully to bind rape-seed enoyl reductase and was a key step in its isolation. Figure 1.9 shows a different example, a ligand for pepsin, and the similarity of size and shape between the ligand and the normal substrate. Another example involving lectins and their ability to bind mannose residues on lipases is described in Chapter 5. It should be noted nevertheless that even affinity columns of high specificity and binding strength are rarely able to give a 'one-step' purification and are best combined with a preliminary fractionation and concentration. They will also need a dialysis since

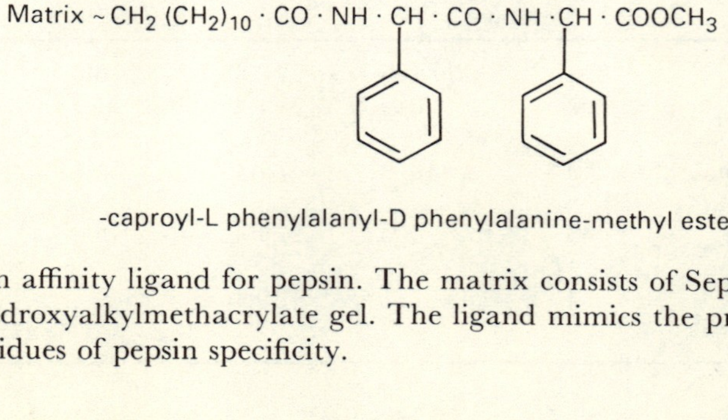

Matrix ~ $CH_2 (CH_2)_{10} \cdot CO \cdot NH \cdot CH \cdot CO \cdot NH \cdot CH \cdot COOCH_3$

-caproyl-L phenylalanyl-D phenylalanine-methyl ester

Fig. 1.9 An affinity ligand for pepsin. The matrix consists of Sepharose H1000, a hydroxyalkylmethacrylate gel. The ligand mimics the preferred aromatic residues of pepsin specificity.

they often require very specific buffer ions, and finally the product will need further purification if only to remove the substrate needed to elute it. The main drawback of this approach is that it needs much prior knowledge of the enzyme which may not be available if it has not previously been isolated. Affinity columns are the best way of routinely isolating enzymes on a commercial scale.

The four techniques described here resolve mixtures by using different properties of the molecule and are therefore independent. In combination, they are capable of isolating almost any enzyme or other protein that can be extracted in solution. It is rarely possible to isolate more than one at a time, and because of the numerous steps involved, yields >10% are exceptional and are often much less. It is impossible to predict in advance which of the many combinations is best, and for repeated preparations optimization can obtain better yields.

Obviously it is necessary to have some means of measuring which fraction contains the protein of interest. The enzyme activity is most often used, and a simple assay is desirable since many will be needed. MCH hydrolyses decanoyl CoA, a reaction that can be followed by simple UV spectrometry. It is not the only thioesterase activity present but is by far the major one and can easily be followed from fraction to fraction during isolation. In the absence of enzyme activity it may be possible to use an antibody or, failing that, high-resolution electrophoresis can in principle be used to trace any protein, though it is laborious.

Table 1.6 summarizes the steps that were used to isolate MCH. There were problems with the source. It was found that the level of MCH in the mammary gland could drop to nearly zero very quickly towards the end of lactation, resulting in unexpectedly low yields. The final step which involved gel filtration also gave difficulties. In the first attempt the enzyme was transferred to a buffer at pH 6.0 by this operation. It immediately aggregated and precipitated from solution and was found to undergo extensive conformational change with formation of inter-chain disulphide bonds at that pH. This behaviour is unique: it is very unexpected for an

Table 1.6 Steps in the isolation of MCH from rat mammary glands. (Adapted from L.J. Libertini and S. Smith (1978). *J. Biol. Chem.* **253**, pp. 1393–1401.)

	Volume (ml)	Protein (mg)	Activity	Specific activity (units/mg)	Recovery (%)
Initial extract	855	17 180	99 200	5.8	100
Ammonium sulphate ppt.	100	2810	61 100	21.1	62
Calcium phosphate gel	260	1378	47 700	31.1	48
DEAE cellulose fractionation	80	100	27 400	274	28
Sephadex gel filtration	3	11.7	12 900	1100	13

Table 1.7 Amino acid codes and the composition of MCH

Amino acid	Three-letter abbreviation	One-letter abbreviation	Residues/mole MCH
Alanine	Ala	A	20.5
Arginine	Arg	R	11.0
Asparagine	Asn	N	25.5
either amine or acid	Asx	B	
Aspartic acid	Asp	D	
Cysteine	Cys	C	2.9
Glutamine	Gln	Q	24.8
either amine or acid	Glx	Z	
Glutamic acid	Glu	E	
Glycine	Gly	G	17.3
Histidine	His	H	9.2
Isoleucine	Ile	I	16.0
Leucine	Leu	L	32.4
Lysine	Lys	K	18.8
Methionine	Met	M	1.3
Phenylalanine	Phe	F	16.5
Proline	Pro	P	15.8
Serine	Ser	S	15.8
Threonine	Thr	T	11.5
Tryptophan	Trp	W	0.8
Tyrosine	Tyr	Y	5.8
Valine	Val	V	12.0

enzyme to show such effects at physiological pH. Inspection of the amino acid composition (Table 1.7) shows a relatively high histidine content. This residue is the only one with a pK near 7 (the pK is the pH at which a group is half ionized) apart from the N-terminus which in MCH turns out to be acetylated. It may be that histidine ionization is connected with the conformational shifts. The preparation had to be repeated this time keeping the pH at all times above 6.5. Events of this kind are not unusual in novel enzyme isolations and explain why they can take so much time to achieve.

Proof of Homogeneity: Electrophoresis
Sequencing work demands the highest possible level of homogeneity in the preparation. During preparation, the specific activity of the enzyme increases to a maximum and then remains constant, as would be expected when homogeneity is reached. This is however quite inadequate to demonstrate lack of contaminants, and electrophoretic analysis is now invariably used to demonstrate purity.

Electrophoresis is the most widely used technique in biochemistry—75% of all papers in the literature contain examples. It has developed many variants, and while this is not the place to go into experimental detail, it is important to understand how they can be used to obtain different kinds of information about the molecule of interest, both proteins and nucleic acids.

Consider a charged molecule in a solution, supplied with a pair of electrodes so that a current can be passed. When the current is switched on, the molecule accelerates in the appropriate direction, but eventually reaches a steady velocity because of viscous drag in the solvent.

$$eV = fm$$

where e is the net charge on the molecule, V is the potential gradient, m is the mobility and f the frictional coefficient, is thus the fundamental electrophoretic equation. Molecules vary in two respects, e and f. The frictional coefficient depends on size and shape—size can easily be allowed for by taking account of the charge density,

$$\frac{\text{Net charge}}{\text{Particle mass}}$$

Thus electrophoretic resolution depends on the charge density and the shape and this can provide extremely high resolving power in favourable cases. One of the best examples is nucleic acid sequencing gels. The main practical problem is one of stabilizing the solutions against gravitational instabilities for the long periods needed for electrophoresis. This has been solved by containing the solvent within a gel structure. Note however that there are two quite distinct types of gel electrophoresis—those like nucleic acid sequencing in dilute agarose where the mobilities are the same as those in free solution and the pores are far too large to give a filtration effect, and concentrated gels. These, nowadays almost always polyacrylamide, are made so that the pores have dimensions comparable to protein molecules. Thus mobilities are lower than in free solution and large molecules, for example globular enzymes, migrate more slowly than small ones with the same charge density and shape.

Protein structures are susceptible to the nature of the solvent, and electrophoresis in dissociating solvents has been used for a long time to investigate the subunit structure of enzymes. Concentrated urea or formamide were used, but it has been found that detergents, such as sodium dodecyl sulphate (SDS) are almost universal solvents for proteins, and many which had been almost impossible to handle, such as actomyosin, became accessible in this solvent. It has the disadvantage that proteins are taken down to their subunits, so the intact molecule cannot be used. Another problem swiftly encountered when it was introduced was the rapid breakdown of samples to small peptides—very often within 24 h. This can be avoided by boiling the solutions on making up, and appears to be due to bacterial proteases found on all glassware, and which are activated by the detergent. Although its ability to act as a universal solvent is important, SDS electrophoresis has become widely used for estimating peptide chain molecular weights. Proteins in SDS solution form complexes containing about 40% SDS by weight. They are therefore of very high charge density which is, moreover, constant from one protein to another. Any initial differences in charge density are swamped by the SDS. Highly charged flexible molecules form rod shapes in solution, and the length of the rod will depend on the number of residues that it contains. The mean residue weight of a protein (molecular weight/number of residues) for globular proteins is remarkably constant and is 110. Thus there is a systematic relationship between the length of the molecule and its molecular weight. It is found that the electrophoretic mobility is related to the length of the molecule by

$$\text{Log (molecular weight)} = km$$

There are various theories based on gel filtration, and the connection with gel filtration columns is obvious to account for this empirical observation; none of them is entirely satisfactory mainly because the gel structure is not well understood.

With suitable calibration an estimate of molecular weight can be obtained. In the best cases a precision of about 1% has been claimed, but there are many known anomalies and errors of 10% are not unknown. DNA sequencing gels (see below) work on exactly the same principles as SDS molecular weight gels.

It is possible to estimate molecular weights for multi-subunit proteins in 'non-denaturing gels' by measuring mobility as a function of gel concentration. Electrophoresis in a pH gradient will result in the protein moving from all directions to come to rest at its isoelectric pH, providing the current is in the right direction. This method, isoelectric focusing, has been known for many years but has not been used until recently because ways of making stable pH gradients were not available. They now are, and as a result so-called two-dimensional electrophoresis has been developed. In this, the mixture of proteins is first separated by isoelectric focusing, and then separated again by SDS electrophoresis so that the proteins are distributed over the whole area of a rectangular gel. Proteins have traditionally been detected after separation by using histological staining methods but these are not sensitive enough for two-dimensional gels, where radioautographs were used. Recent development of very sensitive silver staining techniques means that two-dimensional gels can now be used much more freely.

Whenever possible, the very high resolution of the two-dimensional gel should be used to demonstrate homogeneity of samples for sequencing work. A single SDS run may not be sufficiently discriminating. It would, for example, completely fail to reveal polymorphisms leading to charge differences, like the haemoglobin variants. It should at least be supplemented by a non-dissociating run if possible.

Storage

Having isolated the protein it is desirable to store it. The most convenient form is as the de-salted freeze-dried powder, which will involve a gel filtration or dialysis as a final step. Many proteins are insoluble in water and MCH, for example, was also unstable. It is possible to freeze-dry from salt solution but this leaves the solid with very high salt content which may also be destabilizing. Thus many proteins must be stored as frozen solutions. In a recent and very ingenious development, a way of storing solutions unfrozen at low temperature has been devised. Freezing only takes place in the presence of nucleation sites, otherwise the water simply supercools. By making an emulsion, only the water droplets containing ice nucleation points freeze, while the great majority remain liquid. This avoids damage due to freezing and thawing.

Interaction of Yeast FAS and MCH
In the course of investigating yeast as a possible host for the MCH gene it was found that yeast homogenates in the presence of MCH made fatty acids with a shorter chain length than in its absence. This means that there is an interaction of some sort. It was then found that purified yeast FAS immobilized on a column could selectively bind MCH from rat mammary-gland cytosol. The FAS bound

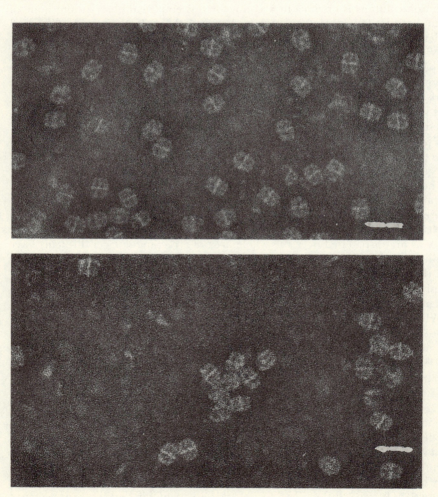

Fig. 1.10 Electron micrographs of yeast fatty acid synthetase (top) and yeast fatty acid synthetase + MCH (bottom). The bar is 50 nm. Tungsten-negative stain on a carbon grid. The individual molecules show an equatorial band and added MCH gives a slight increase of electron density in the polar region.

The Basic Operations of Biotechnology

6 moles MCH per mole, which fits quite well with the presence of six complete sets of FAS enzymes in the molecule, and the strength of binding was considerable—about the same as an antibody–antigen reaction and of comparable specificity. This is a remarkable observation, and it would be dangerous to generalize. The ease of isolation of MCH from the mammary gland suggests that MCH actually binds more strongly to yeast FAS than it does to its natural partner. Figure 1.10 illustrates some electron micrographs which attempt to localize its binding site on the yeast FAS molecule. The yeast FAS is so large that it can easily be visualized in the electron microscope but MCH is relatively small and cannot be seen clearly. Careful scrutiny suggests that it might be located in the polar region. Addition of an anti-MCH antibody brought about clumping. These observations, although the interaction with yeast FAS may well be a freak, suggest that the MCH might be able to function in an alien environment.

STAGE 3. PROBES AND DNA SEQUENCING

Partial Sequencing and Probe Synthesis
It is possible to determine the sequence of amino acids in a peptide by the Edman degradation reaction. The N-terminus of the peptide reacts with phenylisothiocyanate, which makes the peptide link labile. Following removal of the phenylthiohydantoin formed, the next N-terminus can be reacted, and so on. The reactions are shown in Fig. 1.11. Each amino acid residue produces a characteristic derivative, which can be identified by chromatography. Recent developments use fluorescent derivatives to increase the sensitivity of detection but the basic reaction has been in use for 30 years.

In favourable cases, it is possible to complete 20 cycles and thus sequence 20 residues before losses and ambiguities become too great. In practice 10 residues is more common. It is therefore necessary to break the protein into smaller peptides, and fractionate the mixture before sequencing each of them in turn.

The protein must be broken up at least twice, by proteases with different specificities so that the peptides can be arranged in order if the full sequence is wanted. For our purpose this is not needed. With luck, the necessary amino acid sequence might emerge from sequencing of the intact protein from the N-terminus. Attempts to do this with MCH revealed only fractional N-terminal amino acids—of the order of 0.05 mole per mole. So the molecule was broken up with trypsin and the peptides fractionated by methods similar to those used to fractionate proteins. One of the peptides proved to have no free amino N-terminus, and by mass spectrometry was shown to be acetylated on the N-terminal glutamate. This is not uncommon, but makes the task more difficult.

For the limited sequencing needed, it is usual to select the peptide peaks that look reasonably abundant and homogeneous. Unless great care has been taken to purify the protein it is easy to choose a peptide deriving from an impurity at this stage. Especially if it proves to be a good one for probe construction, a great deal of time can be wasted, since the probe will pick out the wrong mRNA and this will not be discovered until DNA sequencing has been completed. Even then it will not be obvious unless a number of peptides have been sequenced. They must all be present in the final sequence derived from the DNA. For this reason alone it is

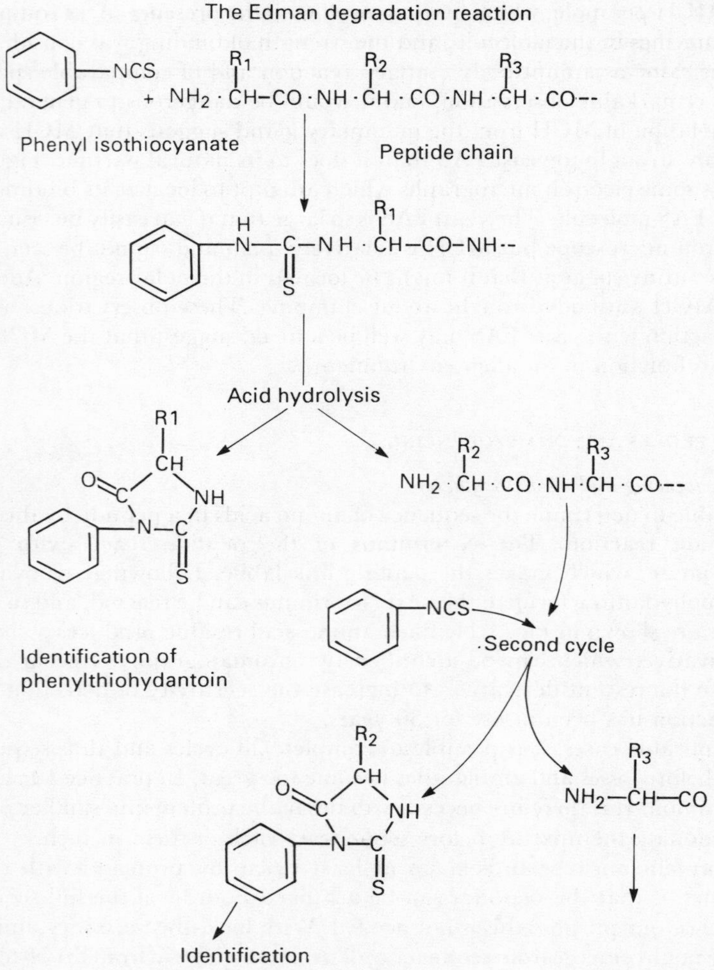

Fig. 1.11 The Edman degradation for amino acid sequencing.

desirable to obtain about 15–20% of the sequence by analysis of the peptides, although in principle a run of only six amino acids is needed.

Table 1.7 shows the amino-acid composition of MCH and the one-letter and three-letter codes used for amino acid residues. The one-letter code is now coming into widespread use because it is much more convenient for computerized data handling.

Table 1.8 shows some peptide sequences obtained from MCH, and the corresponding nucleotide probes based on these sequences. The genetic code has to be used to devise these (Table 1.9) and as is well known, it is redundant since several triplets code for the same amino acid in many cases. Thus in order to avoid

The Basic Operations of Biotechnology

Table 1.8 MCH peptides and the corresponding DNA probes. Where multiple choices are shown all possible combinations are made

Peptide	Met– M	Gln– Q	Pro– P	Asp– D	Arg– R
Code (RNA)	AUG	CAA	CCU	GAU	GCU
		G	C	C	C
			A		A
			G		G
Complementary probes (DNA)	TAC	GTT	GGA	CTA	CGA
		C	T	G	G
			C		T
			G		C
Peptide	M	E	P	L	H
	AUG	GAA	GGA	UUA	CAU
		G	C	CUU	C
			U		
			G		
DNA probe	TAC	CTC	CCT	AAT	GT
		T	G	G	C
			A	A	
			C	G	
Peptide	F	I	F	D	K P
DNA probe	AAA	TAG	AAG	CTG	TTT GG
	G	A	A	A	C

having to synthesize large numbers of possible probes one seeks peptides containing the amino acids with unique codes. These like tryptophan and methionine are relatively uncommon. It is not advisable to rely on a single probe, and ideally one seeks probes from opposite ends of the molecule. Isolated mRNA which reacts with both has a better chance of containing the whole sequence.

Probe Construction
The idea of probes arises from the double-stranded nature of DNA itself. If one takes a string of nucleotides, for example ATTGCGT, then because T is complementary to A and G to C this string would interact strongly, by hydrogen bonding and goodness of fit due to the shape of the residues, with the string TAACGCA. It is sufficiently discriminating to find this string only, and would react much less strongly with a string with only one residue different. A represents adenosine, T thymidine, C cytosine and G guanidine. (Note that RNA, apart from the different pentose also contains uracil U, instead of thymidine.) Thus if a probe is added to single-stranded DNA, there will be a collection of non-specific interactions which can be eliminated by raising the temperature, leaving the probe still attached to its complementary sequence. The temperature at which the probe dissociates depends on its length and the goodness of match. In practice a match of about 18 nucleotides gives a useful increase in stability and provides sufficient discrimination to avoid false-positives. This means that a sequence of at least six amino acids

Table 1.9 The genetic code*

Amino acid	Codon	Amino acid	Codon
F	UUU	P	CCU
	UUC		CCC
L	UUA		CCA
	UUG		CCG
	CUU	T	ACU
	CUC		ACC
	CUA		ACA
	CUG		ACG
I	AUU	A	GCU
	AUC		GCC
	AUA		GCA
M	AUG		GCG
V	GUU	Y	UAU
	GUC		UAC
	GUA	H	CAU
	GUG		CAC
G	UCU	Q	CAA
	UCC		CAG
	UCA	N	AAU
	UCG		AAC
	AGU	K	AAA
	AGC		AAG
D	GAU	E	GAA
	GAC		GAG
C	UGU	W	UGG
	UGC		
R	CGU		
	CGC	G	GGU
	CGA		GGC
	CGG		GGA
	AGA		GGG
	AGG		
		Stop	TGA
			TAA
			TAG

*Note: This is the RNA version; for DNA, U should be replaced by T, as in the stop codons shown.

is desirable. Reference to Table 1.9 shows that because the code is redundant, it will always be necessary to synthesize a mixture of nucleotides to act as a probe. The number can be minimized by choosing the appropriate amino acids, such as methionine or tryptophan, if one is so fortunate as to have an appropriate peptide sequence. Also codon frequency varies considerably: some are so rarely used that they could be omitted with small risk. Nowadays since nucleotides are relatively easy to synthesize, this chance is not taken. Radioactive nucleotide probes are available to order commercially.

The Basic Operations of Biotechnology

Table 1.8 shows some peptides that were used to make probes for MCH, and some others that were not used but will be needed later (see Fig. 1.13).

Isolation of mRNA and Synthesis of cDNA

Ultimately we need a piece of DNA coding for MCH for insertion into a host. In principle once a probe is available one could search for the gene in the nuclear DNA. This is a viable approach for a simple organism like *E. coli* but it is very difficult to find a gene in the mass of DNA present in mammalian cells. It is easier to isolate the mRNA and then by using RNA directed DNA reverse transcriptase make the DNA complementary to it—cDNA.

mRNA is isolated from tissues by density-gradient centrifugation, and a rough size classification can also be made. The molecular weight of the protein gives a minimum estimate of the mRNA size. There were problems with MCH because the tissue produces large amounts of casein with an mRNA similar in size to the MCH. Another method is to use an antibody to the protein, to precipitate it while it is still attached to the ribosome on which it is synthesized, together with the mRNA.

Having isolated the mRNA it is then checked to see that, in a cell-free protein-synthesizing system, usually rabbit reticulocytes, it can indeed direct the synthesis of the correct protein. This is identified by electrophoresis and antibody reaction. Once confirmation that the mixture contains the correct mRNA has been gained, then it is used as a template for cDNA synthesis. If this is successful it forms a single-stranded DNA which is then converted to the double-stranded form by one of the other key enzymes of genetic manipulation, DNA polymerase.

Most of the DNA will not be the one we seek, so it is now cloned. This operation will increase the amount of DNA but also allow us to use our probe to find the MCH gene.

The cDNA was incorporated into a plasmid, *pBR322*, commonly used and which contains a gene-conferring resistance to tetracycline, *E. coli* infected with the plasmid and 6000 tetracycline-resistant colonies established. This number was chosen on the basis of the likely frequency of the MCH cDNA in the mixture, since most of the cDNA will be for casein. Finally, the nucleotide probe was used to identify the colonies that contain the MCH cDNA. In fact two probes were used, one finding 18 clones and the other 16, of which 6 were in common with the first. Choosing the clone which contained the longest piece of cDNA, this was then multiplied to give a supply of single-stranded DNA once more. Clearly this piece of DNA can itself act as a probe for mRNA, and as a check it was shown that it did correctly isolate the MCH mRNA. From its size there was a good chance that it would contain most, if not all, of the sequence for MCH.

DNA Sequencing

The next step is to sequence the DNA and from this deduce the complete amino acid sequence of the protein. Sanger's dideoxy method is the most widely used. It begins with yet another cloning step.

The M13 phage exists as a single circular strand, but when it infects *E. coli* it makes a second complementary strand to form the replicative form. This then produces large numbers of the single-stranded form which are secreted into the medium and can be isolated. cDNA can be incorporated into the phage, and multiplied in this way, and special varieties are available with restriction sites that make incorporation easy, as well as a galactosidase gene that aids selection. The mechanism of replication is such that the added DNA may be transcribed in either direction.

The sequencing reaction uses a modified DNA polymerase, which is able to synthesize the complementary strand from added nucleotides by extension of a primer string. This is chosen to be complementary to the appropriate spot on the M13 phage so that extension goes through the region containing the sequence of interest. Sequence information is obtained by replacing a proportion of the added nucleotide with the dideoxyribose analogue, so that when this residue is incorporated it prevents further elongation. The result is a mixture of nucleotides

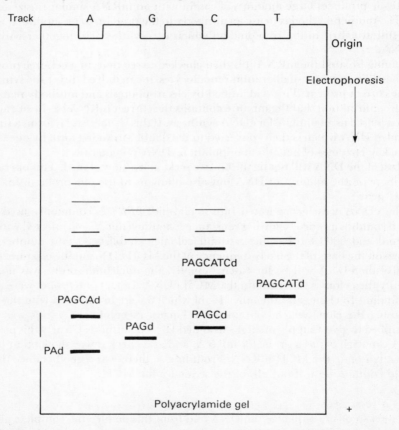

Fig. 1.12 Part of a sequence gel for DNA. The mobility is closely related to the length of the chain. P represents the primer sequence: the four tracks contain each, one of the four different dideoxy nucleotides, and therefore indicate the relative positions in the chain. The sequence is AGCATC.

of varying length. The reaction is performed four times replacing each nucleotide in turn, and the products separated by gel electrophoresis. As has already been explained for SDS gels, the mobility is strictly related to the length of the molecule, and it is thus possible to read the sequence from the relative positions of the zones on the gel. A diagrammatic result is shown in Fig. 1.12. Because of the small amounts, radioactive derivatives are used to aid detection, though fluorescent markers are now coming into use, together with automation of the technique. It is possible to obtain about 300–400 residues from each gel and a sequence like MCH needs several determinations with overlapping fragments to assemble the full sequence.

Some caution is needed and the method is not error-free. Clerical errors in reading gels and recording data average 1 residue in 500 and there can be difficulty with certain sequences. Data handling is invariably by computer, and translation of a DNA sequence obtained in this way produces six amino acid sequences. This is because the sequence can be read in either direction, starting at any point. It is then necessary to search the sequence for known peptides. This, if it is successful, quickly establishes the correct direction, but even then omission of a single residue will produce a frame shift, and it is usually necessary to repeat some parts of the sequence. The more peptide data available the better at this stage. Another problem arises in establishing the beginning and end of the sequence. The N-terminus is always a methionine, and for many proteins, unlike MCH, the sequence for this part will be known. The C-terminus should coincide with a stop codon, but it is important to compare the deduced amino acid composition and number of residues with that of the protein. Two values, both from SDS gels, were available for MCH, 32 000 and 29 000, corresponding to 290 and 260 residues—a difference of 30 amino acids! Alas, molecular-weight measurement is not precise enough to define the number of residues in a useful way. Figure 1.13 gives the final sequence, the deduced amino acid sequence, and the positions of the known peptides. It has 263 residues corresponding to a molecular weight of 29 300, and a stop codon occurs in the expected place. Comparison with the amino acid composition given in Table 1.7 shows reasonable agreement: amino acid analyses are, like molecular weights, not quite precise enough to be really useful. Thus it appears that MCH is not processed other than by having an acetyl group attached to its N-terminus. It is often found that the outcome of DNA sequencing has extra peptides compared with the isolated protein. Our next example, thaumatin, has them at each end (see below).

STAGE 4. VECTORS AND HOSTS

Transfer to the Host Organism
We now have a piece of DNA capable of coding for the production of our enzyme MCH. However we should note that it is probably not of the same structure as the gene in the original organism. In most non-bacterial organisms (i.e. eukaryotes) the DNA consists of segments called introns, which are expressed and exons which are not, and may occur in the middle of a coding region. A gene will have the following structure:

```
              intron     exon      intron       exon
    DNA       -------   -------   ---------   -------  ----
    nuclear              transcription
    RNA       -------   -------   ---------   -------  ----
              processing and excision of exons in nucleus
    mRNA      intron              intron               intron
              ----------          --------             ----------
```

Thus by copying the mRNA to get our DNA we avoid any possible problems with exon regions, which the potential host might not be able to process properly. Bacteria cannot.

As was made clear when describing DNA sequencing techniques it is possible to use the enzymes of DNA metabolism to insert DNA into pre-existing strands in a controlled way. This uses, in particular, enzymes known as restriction nucleases. They are found in microorganisms, and appear to be part of a defence mechanism against virus attack. They are called restriction enzymes because they restrict the length of the DNA and their great value is that they are highly specific for the nucleotide strings at which they split the DNA chain. About 40 different enzymes are in common use in DNA manipulation.

Bacteria apart from the main DNA contain small pieces of circular DNA known as plasmids. They are strain-specific and are maintained in the population by the normal processes of multiplication—'cloning'—in the case of bacteria and their DNA can use the ordinary protein synthetic mechanisms of the cell. In *E. coli* the cells can be made to take up plasmids by treatment with cold calcium chloride and many well-characterized plasmids are known. Thus, plasmids are an excellent vector system since DNA incorporated into them can be inserted, propagated, and expressed. All the initial work, and many laboratory procedures such as the M13 cloning for DNA sequencing already described were performed in *E. coli* for this reason. However, it is very unlikely that anything produced in *E. coli* will ever be used for food manufacture since it is not an acceptable organism. Indeed with the possible exception of lactobacilli, and *Bacillus subtilis*, it is doubtful if many genetically manipulated bacterial products would be acceptable to the regulators.

A target for MCH is amongst others, a flowering plant, such as rape seed, but it has been incorporated into yeast, using a plasmid found in that organism. Although it may be expressed it seems not to function. It has been successfully incorporated into animal cells (mouse fibroblasts) kept in tissue culture, and does influence the general fatty acid pattern in the expected way. This is shown by the results in Table 1.10 where the average chain length was shifted downwards. This is the first example of an alteration in fatty acid patterns by gene transfer. The turnover rate of the triglycerides in the transformed cells increased considerably, suggesting that the whole of that part of the metabolism was perturbed. At present we cannot take the MCH development any further with respect to plants. Instead we will consider some work on the insertion of a pea legumin gene into the tobacco plant, as an example for flowering plants as hosts. Of the two objectives of transformation—either to produce an organism with an altered metabolic pattern, so as to influence its content of, for example lipids, or to act as a source of the expressed protein for other uses—MCH is clearly in the former category. So far there is no successful example of this, though it appears to be merely a matter of

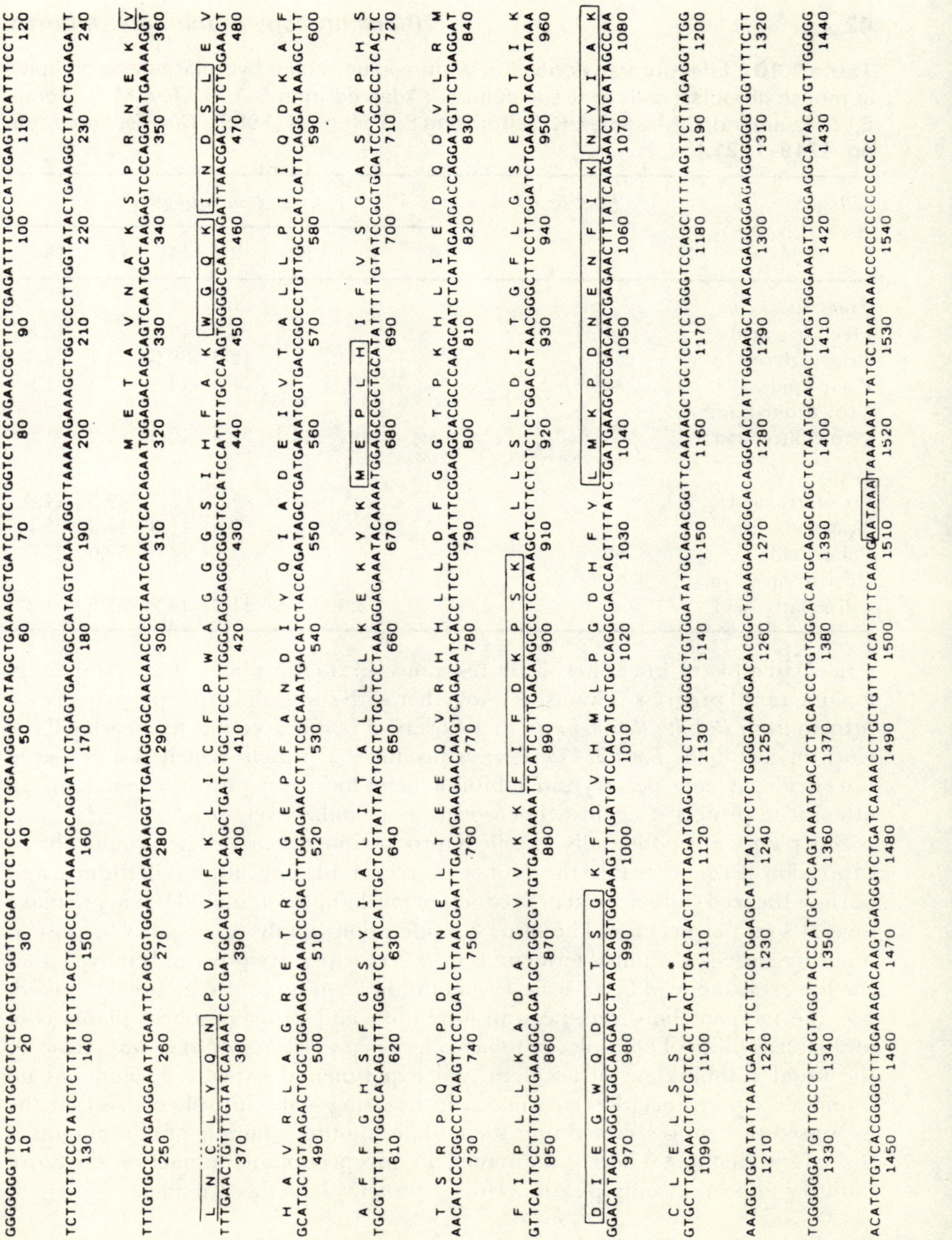

Fig. 1.13 The nucleotide sequence and the deduced amino acid sequence for MCH. The boxed peptides are those expected from direct amino acid sequencing. (From R. Safford, J. de Silva, C. Lucas, J.H. Windust, J. Shedden, C.M. James, C.M. Sidebottom, A.R. Slabas, M.P. Tombs and S.G. Hughes (1987). *Biochemistry* **26**, pp. 1358–1364.)

Table 1.10 Effect of transformation with medium-chain hydrolase gene on lipids in mouse fibroblast cells in tissue culture. (Adapted from S.A. Bayley, M.T. Moran, E.W. Hammond, C.M. James, R. Safford and S.G. Hughes (1988). *Bio-Technology* **6**, pp. 1219–1221.)

Cells	Total lipid (%)	Fatty acid chain length (%)						
		6	8	10	12	14	16	18
Transformed cells								
Free fatty acid	1	–	–	3.0	5.0	17.5	25.7	48.8
Triacylglycerol	57	–	1.4	1.7	12.5	30.9	25.1	28.4
Polar lipids	42	–	–	–	1.3	14.5	32.3	51.9
Growth medium:								
free fatty acid	–	1.4	13.1	40.1	31.5	10.5	2.2	1.2
Control cells								
Free fatty acid	1	–	–	–	–	5.7	49.5	44.8
Triacylglycerol	31	–	–	–	2.6	11.8	57.9	27.7
Polar lipids	68	–	–	–	–	4.0	69.0	27.0
Growth medium:								
free fatty acid	–	–	1.4	5.6	11.9	14.2	62.6	4.3

time. Attempts to insert herbicide resistance into crop plants, for example are making rapid progress. This aspect, together with successful attempts to express a protein from *Bacillus thuringiensis*, in a variety of plants accounts for nearly all the work in this field. *Bacillus thuringiensis* produces a protein which is a powerful insecticide. A cow pea trypsin inhibitor gene has been put into maize, in an attempt to protect it against root worm, in a similar exercise.

But plants are potentially excellent production systems. If we could obtain expression of a gene (to take the most obvious example, say, insulin) in the pea, and harness the seed storage protein production mechanisms, it should be easy to make several tons per hectare. The same considerations apply to enzymes at present made by large-scale fungal fermentation, a very expensive process. It may be that the life expectancy of large-scale fermentation for protein production is limited. For the moment however the transformation and propagation of plants is of limited feasibility. The choice of tobacco (*Nicotiana plumbagifolia*) as host arose by the usual combination of accident and experimental expediency and has no economic significance. Recent success in inserting genes into sheep, so that the expressed protein is secreted into the milk, is another example of non-microbial protein production. There is no immediate prospect of any impact on the dairy industry and so far only pharmaceutical proteins have been made.

The Use of Yeasts: Thaumatin
For the time being therefore the production of proteins for food use is likely to be in acceptable fungi, especially yeasts because some good vectors are available. As an example of this we will consider thaumatin. Thaumatin is a sweet protein (see Chapter 2) which is needed in quantities greater than are likely to be available

The Basic Operations of Biotechnology

from its normal source, a West African shrub (*Thaumatococcus daniellii*), as a food ingredient. It is fairly easy to isolate, though since it is not an enzyme it had to be detected by a combination of electrophoresis and specific antibodies. There are five closely related thaumatins. This would not be surprising in view of the widespread population polymorphisms found in all organisms, though in this case all five forms were found in a single plant. Plant samples from a large number of individuals are almost always certain to contain polymorphic forms differing usually by a single amino acid substitution.

The mRNA for one of the forms was isolated by electrophoresis and after selection on the basis of its expected size, it was confirmed that it correctly expressed thaumatin in a rabbit reticulocyte cell-free system. At least half the protein in the tissue chosen was thaumatin, so the mRNA was abundant, unlike the MCH case discussed above.

The corresponding cDNA was then made, and sequenced showing that thaumatin is synthesized in a pre–pro form. The isolated thaumatin has 203 residues, but the mRNA codes for the following,

```
         pre          thaumatin              pro
NH  ----------   ----------------  ----------  ---COOH
         22             203                     6
```

and while a hydrophobic pre-chain is common, and is removed after synthesis very often as part of a membrane transit process, a hydrophilic C-terminal sequence is very unusual and was previously known only in interferons. The cDNA coding for pre-prothaumatin, prothaumatin, thaumatin and prethaumatin was then made and separately inserted into *E. coli* via a *pBR322* plasmid. All the forms were made by the cells but in poor yield, and as expected *E. coli* did not carry out any post-synthetic processing. The yield was low despite the use of promoters. When DNA is transcribed to make RNA, the frequency of transcription is controlled by promoter regions of the DNA, found upstream of the coding region. Such pieces of DNA have been isolated and incorporated into plasmids used as vectors, where they greatly increase the rate of synthesis of the desired protein. One widely used is that for galactose breakdown. This and others like it have a further advantage that they can be switched on and off by adding inducers to the culture medium. Thus a potentially lethal gene product can be left unmade until the population has grown to its maximum, and then the synthesis switched on. Other switches, dependent on temperature change or even NaCl concentration, are known.

The next step was to use *Saccharomyces cerevisiae* and *Kluyveromyces lactis*, since these are both acceptable food organisms. For this a promoter which regulates the synthesis of glyceraldehyde-3-phosphate dehydrogenase was identified and incorporated into the vector. A shuttle vector was used, which can operate in both *E. coli* and yeasts. The general structure of such a vector is shown in Fig. 1.14. Shuttle vectors have been used widely in laboratory work to transform yeasts. Both the yeasts made thaumatin when provided with cDNA coding for pre-prothaumatin in reasonable yield, but with only poor yield from prothaumatin. Thus yeast seems to require the pre-signal sequence, and can properly process it. However the

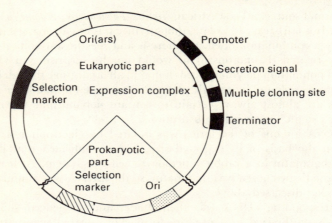

Fig. 1.14 A shuttle vector for eukaryote cloning. The eukaryote and prokaryote parts are shown, and permit cloning either in *E. coli* or in yeast. Each contains a selection marker, such as chloramphenicol resistance and an autonomous replication signal for host DNA polymerase. In addition, a multiple cloning site and promoter are in the eukaryotic part. (Adapted from K. Esser and J. Kamper (1988). *Process Biochemistry* **23**, pp. 36–41.)

improved yield was unexpected since the bovine chymosin gene gave a better yield in the absence of the pre-signal.

It will be seen (Fig. 1.14) that the vectors include signals for secretion. Yeast secretes few proteins, and does it in the same way as animal cells. However, attempts to use the appropriate signals derived from animal DNA did not give very good results. This difficulty has been resolved by finding the signals used in yeasts themselves. Yeast can glycosylate after synthesis, but generally does not do so for inserted cDNA, and it does not always carry out processing operations correctly. It cannot usually process nuclear RNA, so DNA devoid of introns must be used.

Clearly there remains much to be learned about the use of yeast as a production system. Even so commercial manufacture of chymosin in yeast is a reality.

Plants as Hosts
Plants can be cloned. Some of them are usually propagated in this way simply as cuttings, but cloning from single cells or small clumps of cells, permitting selection and multiplication more like microorganisms, is also possible. Even so the numbers do not approach the countless millions of bacteria or yeast cultures. In one of the most highly developed examples, clonal oil palm, a single plant may give rise to 100 000 via tissue-culture methods. These numbers are much too small to permit the sort of low-probability selection methods feasible with *E. coli*, while the propagation itself is far more difficult. In some of the earliest successful attempts to produce transgenic plants the actual number of individual plants grown containing the foreign DNA was <10: the contrast with microorganisms is obvious.

The Basic Operations of Biotechnology

It follows that insertion of DNA must be correspondingly efficient, and there is little scope for methods that insert a large number of fragments in the hope that one of them is complete and correct. Two approaches have been used. The first modelled on methods for microorganisms uses plasmids contained in *Agrobacterium tumefaciens*. This organism infects dicotyledons, and causes tumours to appear due to the presence of tumour-inducing plasmids (*Ti*). It is inserted either by simply exposing shoots or roots to the bacterium, or by using protoplasts. These are cells which have had their cell walls removed by treatment with cellulases. Some yeast mutants lack cell walls, and are also used as protoplasts. The absence of a cell wall leads to a higher degree of infection.

The second approach consists of the direct insertion of DNA into individual cells. DNA inserted in this way is, surprisingly enough, incorporated into the genome and transmitted to progeny. Micro-injection has been used, and a technique called 'electroporation' is sometimes effective. In this, high-voltage d.c. current is applied to the cells and apparently causes transient holes in the membrane, permitting ingress of DNA. In another method, small tungsten particles were coated with DNA and literally fired into onion cells with a gun. These techniques, usually aimed at monocots, seem to work albeit lacking in the elegance of vector construction usually found in transformation techniques.

The main remaining difficulties are that *Ti* plasmid-based methods are not effective on cereals, while regeneration of plants from transformed cells is never easy and is especially hard in cereals, so the main food crops are the most difficult to work with. Another aspect of transformation is to find ways of ensuring that the inserted gene is expressed only in the desired tissue. This is why our next example is interesting: it proved possible to obtain expression of genomic DNA for legumin, a storage protein of the pea, not only in the tobacco plant but specifically in the seeds of the tobacco plant.

Examination of the genes for the major seed storage proteins of legumes showed they had a highly conserved sequence upstream of the coding region, which was thought to be a candidate for a tissue-specifying regulatory element. (Note that this is a situation where only genomic DNA can be used; there would be no point in trying to use cDNA for this work.) Accordingly, the DNA was incorporated into an *Agrobacterium* plasmid. This is a highly developed vector containing marker genes for nopaline synthesis, selection genes for kanamycin resistance, and the ability to grow in both *E. coli* and *Agrobacterium*. Also, since there is little point in transforming a plant only to see it produce tumours, the oncogene part is deleted.

After infection of leaf tissue, new plants were regenerated from shootlets which grow on the edges of the cut leaves. Eventually these plants flowered and the seeds could be examined. Some, but not all the transformed plants had pea legumin in the seeds and nowhere else.

The mature protein was produced, showing that the tobacco had been able to excise three introns in the gene, and carry out post-translational proteolytic cleavage of the legumin chain. Similar observations have been made for soy glycinin in petunias, and phaseolin in tobacco. Most recently, the ACP from spinach has been inserted into *Brassica napus*, rape seed, using the regulator segment for napin, the storage protein, to bring about expression in the seed. The effect of this on the triglycerides has so far not been revealed. It is possible to transform and regenerate rape fairly easily, and this may give it an advantage over soy beans. The breeding of new varieties of rape seed, with an even wider range of fatty acid profiles is now in progress. Classical breeding methods have by no means been abandoned and will almost certainly produce results before transformed plants go into large-scale manufacture.

Patents
In a survey of industrial biotechnology one has to mention patents. A number of these now exist on basic biotechnology techniques, though their exact scope and enforcement is currently a matter for litigation. Patents confer a right to prevent other people from operating the process described, but do not actually confer the right to operate it on the patent holder. Operating it may in its turn infringe other people's patents. What tends to happen in a situation like this is that companies build up portfolios of patents and then trade them for the right to use others. Nevertheless, the existence of patents may have a considerable influence on whether a particular company develops a process or not. They often have a retarding effect since the uncertainty they create is one more reason for doing nothing.

Conclusions

Contemporary food industry biotechnology is still at the stage of testing the feasibility of undertakings, which if successful will have very far-reaching effects. This is the case with work on the use of plants as hosts, much of which is going on under conditions of commercial secrecy. One cannot do much more than indicate the general trend of such developments. They will not only bring about some changes in the origin and nature of food ingredients, but have implications for other branches of biotechnology. These developments are likely to happen in the foreseeable future. The use of yeasts as sources of enzymes for food use is gathering pace, and is on a somewhat nearer time horizon. There are many applications of enzymes in food processing which will form the subject of the following chapters. There are not many examples yet of the use of enzymes obtained by genetic manipulation, and there is almost certainly scope for improved performance by introducing them. Even so, it is now evident that these ventures are not easy, certainly not cheap and are by no means guaranteed a successful outcome. One conclusion is perhaps a surprise, but a major constraint is lack of knowledge of the basic biochemistry of plants. Even where pathways are known, little information is available on the enzymes responsible.

Chapter 2

Sweeteners

Introduction

In the fourteenth century in England the price of a pound of sugar would have bought 29 lb of butter or 400 eggs. Its use was confined to wealthy households, and it did not become relatively less expensive until the sea routes to the East were opened up in the late seventeenth century. Sugar-cane has been known in India for at least 2000 years. Even then it was not particularly cheap, since a pound cost the same as a quart of claret! Consumption has increased enormously, and in the UK the annual per capita intake of sugar is now about 40 kg, one of the highest rates in the world (which averages about 20 kg per capita). In China the average is only 7 kg, but this is not due to lack of demand. Eskimos lack the enzyme sucrase which breaks down sucrose during digestion, and were never exposed to it until recently. More than 90% of the world's production is under quantity and price control for political reasons, but sucrose is nevertheless a very important commodity in international trade. Obviously the flavour provided by sucrose is seen by the consumer as very desirable, and sucrose is almost unique in being a pure chemical sold in bulk directly to the shopper. The trend in recent years has been for the supermarket customer to buy less and less ingredients (and flour is one such), and more and more to replace these with already made articles. This is all part of a trend, which includes washing machines and vacuum cleaners, to minimize the labour of caring for a home and family. But sucrose seems to avoid this perhaps because one of its main uses is by addition to drinks such as tea and coffee. Attempts to sell pre-sweetened tea and coffee have not succeeded. It is used by food manufacturers on a large scale, in soft drinks, confectionery and baked products, as well as in less obvious ones like baked beans in cans, tomato sauce and of course canned fruit.

Although it is still by far the most dominant sweetener, sucrose has been challenged by a variety of other sweetening agents. There are four reasons for this. First, experience during war-time scarcity showed that there was a correlation between the level of sucrose in the diet and the incidence of dental caries. Secondly, a number of nutritionists have claimed that sucrose may be too large a part of the diet in developed countries, and as part of a general attempt to reduce obesity, sucrose intake should be cut. Thirdly, about 1 in 50 of the European population has some degree of diabetes, which requires them to exercise control over their carbohydrate and fat intake. This is usually done by avoiding sucrose, amongst other things. The fourth reason is that, despite its dominant position, new bulk sweeteners, the high-fructose syrups, have been developed which offer a commercial challenge and have replaced sucrose in major applications, at least partly because they are cheaper.

Sucrose, although the dominant sweetener, is by no means the sweetest substance known. Table 2.1 lists a variety of sweeteners, which will be the subject of this chapter, and their relative sweetness, taking sucrose as the standard.

Sucrose

Sucrose, because it does not actually provide very intense sweetness can be used at high levels to give a viscosity needed in some products. It is also widely used as a bacteriostat since at relatively high concentrations bacteria cannot grow in it. The precise mechanism of this effect is unclear and is usually expressed for quality-control purposes as a combination of solution vapour pressure described as a 'water activity' and the pH, on which the preservative effect also depends. Jams and numerous confectionery products, as well as some preserved dried fruits, and even a few meat products (mincemeat is the only familiar European one, but otherwise mainly Arabian dishes) depend for their shelf-life on high sugar concentrations.

Sucrose is the main form in which carbohydrates are translocated in plants, and occurs widely. Table 2.2 lists some mono- and disaccharide concentrations in common fruits and other consumables, while Table 2.3 contains a comprehensive list of food beans and seeds which equally contain sucrose, as well as some

Table 2.1 Relative sweetness in comparison with sucrose on an equal weight basis*

Sucrose	1	Steviolbioside	100
Fructose	1.3	Rebaudioside A, B	300
Glucose	0.7	Saccharin	300
Galactose	0.3	Aspartame	180
Maltose	0.3	Acesulfame K	150
Lactose	0.5	Thaumatin	2000
Lactulose	0.5	Morellin	3000
Glycyrrhizin	30	Cyclamate	30

*The numbers are an indication only: the perceived sweetness varies with concentration and other peripheral circumstances.

Table 2.2 Monosaccharides and disaccharides in fruits and consumables. (Adapted from C. Bucke (1979) in *Developments in Sweeteners*, **1**, 43.)

Commodity	Glucose (%)	Fructose (%)	Sucrose (%)	Maltose (%)
Apple	1.17	6.04	3.78	trace
Juice	3.1	6.4	1.1	
Apricot	1.73	1.28	5.84	
Blackberry	2.48	2.15	0.59	0.66
Blueberry	3.76	3.82	0.19	0.08
Currant	3.33	3.68	0.95	0.64
Gooseberry	3.29	3.90	1.21	
Grape (*Vitis labruscana*)	6.86	7.84	2.25	1.58
Grape (*Vitis vinifera*)	5.35	5.33	1.32	2.19
Orange juice	2.2	3.0	5.2	
Peach	0.91	1.18	6.92	0.12
Pear	0.95	6.77	1.61	0.31
Plum	3.49	1.53	4.94	0.15
Raspberry (red)	2.40	1.58	3.68	
Raspberry (black)	4.56	4.84	1.90	
Cherry (sour)	4.30	3.28	0.40	
Cherry (sweet)	6.49	7.38	0.22	
Strawberry	2.09	2.40	1.03	0.07
Tomato juice	1.2	1.7	0	
Beet	0.18	0.16	6.11	
Carrot	0.85	0.85	4.24	
Cucumber	0.86	0.86	0.06	
Lettuce	0.25	0.46	0.10	
Melon (honeydew)	2.56	2.62	5.86	
Rhubarb	0.42	0.39	0.09	
Tomato	1.12	1.34	0.01	

trisaccharides. While these compilations make it clear that simple carbohydrates are abundant, the relative proportions vary considerably from species to species. It should also be made clear that the figures given are representative, but will certainly vary from variety to variety, and from season to season.

The relative abundance of mono- and disaccharides in different species is under the control of enzyme systems, in much the same way as the different fatty acids discussed in Chapter 1, and is as little understood. Given that varieties of cane or beet rich in sucrose and which, between them, can be grown anywhere are already available, there appears to be no commercial demand for alteration of the balance in any particular species.

Sugar-cane (*Saccharum officinarum*) and sugar-beet (*Beta vulgaris* var. *rapa*) are the two commercial sources and form the basis of a highly organized and efficient industry. A few enzymes are used in production. In Japan and parts of the USA, excessive raffinose appears in beet (its presence can always be used to differentiate cane- from beet-sugar, since the former does not contain it) and is hydrolysed with

Table 2.3. Food beans and seeds: sugar content. (Adapted from T.M. Kuo, J.F. Van Middlesworth and W.J. Wolf (1988) *J. Agric. Food Chem.* **36**, p. 34.)*

Seed	Raffinose sugars				Sucrose	Total sugar
	Verbascose	Stachyose	Raffinose	Total		
Malvaceae						
Cotton (*Gossypium herbaceum*) cv. 'Deltapine 61'	tr	23.6	69.1	92.7	16.4	109.1
Leguminosae						
Garden pea (*Pisum sativum*) cv. 'Little Marvel'	19.1	32.3	11.6	63.0	62.3	125.3
Soybean (*Glycine max*) cv. 'Williams 82'	tr	43.4	12.6	56.0	64.2	120.2
cv. 'Amsoy 71'	tr	41.0	11.6	52.6	72.7	125.3
Cowpea (*Vigna sinensis*)	3.6	46.4	3.7	53.7	25.9	79.6
Alfalfa (*Medicago sativa*)	tr	39.5	13.5	53.0	22.7	75.7
Mung bean (*Phaseolus aureus*) cv. 'Berken'	26.6	16.7	3.9	47.2	13.9	61.1
Lima bean (*Phaseolus limensis*) cv. 'Fordhook'	tr	30.3	6.9	37.2	36.0	73.2
Green bean (*Phaseolus vulgaris*) cv. 'Top Crop'	tr	34.3	2.5	36.8	19.4	56.2
cv. 'Blue Lake'	tr	28.1	2.2	30.3	29.0	59.3
Red kidney bean (*Phaseolus vulgaris*)	tr	31.6	3.1	34.7	21.5	56.2
Pole bean (*Phaseolus vulgaris*) cv. 'Kentucky Wonder'	tr	26.2	4.3	30.5	26.7	57.2
Broad bean (*Vicia faba*)	11.4	10.7	2.3	24.4	20.7	45.1
Peanut (*Arachis hypogaea*) cv. 'Florunner'	tr	9.9	3.3	13.2	81.0	94.2
Compositae						
Sunflower (*Helianthus annuus*)	—	1.4	30.9	32.3	65.0	97.3
Safflower (*Carthamus tinctorius*)	—	—	5.2	5.2	18.6	23.8
Cucurbitaceae						
Pumpkin (*Cucurbita pepa*)						

Species						
Cucurbitaceae						
Cucumber (*Cucumis sativus*) cv. 'Marketeer'	—	10.8	9.2	20.0	15.2	36.2
Squash (*Cucurbita maxima*) cv. 'Table King'	—	11.8	7.8	19.6	34.4	54.0
Solanaceae						
Tobacco (*Nicotiana tabacum*) cv. 'Dpl'	—	tr	7.3	7.3	26.8	34.1
Gramineae						
Barley (*Hordeum vulgare*) cv. 'Himalaya'	—	tr	7.9	7.9	14.2	22.1
cv. 'Steptoe'	—	tr	6.3	6.3	11.8	18.1
Triticale cv. 'Fas Gro 204'	—	tr	7.2	7.2	8.4	15.6
Rye (*Secale cereale*) cv. 'Balbo'	—	tr	7.1	7.1	11.5	18.6
Wheat (*Triticum aestivum*) cv. 'Chinese Spring'	—	tr	7.0	7.0	13.8	20.8
Corn (*Zea mays*) P-3737 hybrid	—	—	3.1	3.1	14.2	17.3
OH 43 inbred	—	—	2.1	2.1	15.0	17.1
Oat (*Avena sativa*) cv. 'Dal'	—	tr	2.6	2.6	8.8	11.4
Sorghum (*Sorghum vulgare*) cv. 'WAC 694'	—	tr	tr	tr	8.4	8.4
Rice (*Oryza sativa*) cv. 'Blue Bell'	—	—	—	—	5.6	5.6
Chenopodiaceae						
Spinach (*Spinacia oleracea*) cv. 'Giant Noble'	—	—	4.8	4.8	6.7	11.5
Beet (*Beta vulgaris*) cv. 'Early Egyptian'	—	—	3.7	3.7	3.4	7.1
Euphorbiaceae						
Castor bean (*Ricinus communis*)	tr	1.9	1.9	3.8	35.1	38.9

*Results are in mg/g of defatted meal. tr = trace.

Table 2.4 Sources and actions of some enzymes

Name	EC no.	Source	Action
α-Amylase [(1→4)-αD-glucose glucanohydriolase]	3.2.1.1	Bacteria, e.g. *B. subtilis* or *B. lichinformis*	Split 1→4 links anywhere in chain except near 1→6 branches
α-Amylase [(1→4)-α-D-glucan maltohydrolase]	3.2.1.2	Fungi, e.g. *Aspergillus oryzae*	Removes maltose from non-reducing ends of 1→4 polyglucose chains
Amyloglucosidase (glucoamylase)	3.2.1.3	*Aspergillus* spp.	Removes glucose by 1→4 fission. Can also split 1→6 links, and degrade amylopectin
Amylase Dextrinase [glucan(1→4)-α-glucosidase]	3.2.1.3	Barley *B. cereus*	
Isoamylase [(1→6)-α-D-glucan maltohydrolase]	3.2.1.68	Bacteria, e.g. *Pseudomonas* spp.	Removes maltose from non-reducing ends
Pullulanase [α-dextrin endo-(1→6)-α-glucosidase]	3.2.1.41	Bacteria, fungi	Splits 1→6 links in amylopectin

the aid of a fungal galactosidase to sucrose and galactose. A reactor-based process has been developed.

In cane-sugar production, amylases deriving from *Bacillus licheniformis* are used to degrade starch grains which are carried into the juice during crushing. In damp, high-temperature storage conditions, *Leuconostoc mesenteroides* can make dextrans appear, and dextranases (see Table 2.4) are used to degrade them, as well as galactomannans that can appear in beet extracts. The final stages of sucrose manufacture involve a crystallization step which is inhibited by the presence of other sugars. It is possible that the use of cellulases would aid the extraction of sucrose in the same way as they appear to assist the extraction of many other plant materials.

INVERT SUGAR

In use, sucrose is often converted to 'invert sugar' so-called because during its formation the angle of polarization of light by the solution changes sign. This can be achieved with enzymes—invertase (β-fructofuranosidase, EC 3.2.1.26), usually obtained from yeast, or by acidification. Sucrose hydrolyses to glucose and fructose—invert sugar—relatively easily. The acid content of citrus fruits is sufficient to do this, but the main use of added enzymes is in fancy-chocolate manufacture. Invertase is added to degrade sucrose crystals and form the

Sweeteners

characteristic soft-centre paste. Partial hydrolysis during manufacture is the reason why boiled sweets form a glass rather than crystallize.

There is almost certainly scope for improvement in the specification of enzymes used in sucrose refining so that they are aimed more precisely at the unwanted contaminants. There may also be scope for production of fermentable sucrose in small amounts from unwanted residues. The residues from sugar-cane are well known as the source of rum, following alcoholic fermentation and distillation. In some countries, notably Brazil, sucrose has been fermented to ethanol for use in internal combustion engines. The whole of the land area in the UK could not produce enough ethanol to run the existing number of cars, so this use is likely to be restricted to large countries with few vehicles.

Glucose and Fructose Syrups

The production of glucose is one of the most important examples of the use of enzymes in the food industry, and one of the oldest. Taken together with more recent processes for the conversion of glucose to fructose, it provides the main justification for claims that biotechnology will bring about a new industrial revolution. In the space of 20 years the position of sucrose, which must have appeared impregnable, has been challenged by novel processes. Recently, the world's largest sucrose processor and merchant has bought one of the largest producers of high-fructose syrups, no doubt after consideration of the commercial realities of the sweetener markets of the world.

Commercial glucose is invariably made from starch by acid or enzyme hydrolysis or a combination of the two. Fructose has been made from sucrose as part of invert sugar, from glucose by alkali-catalysed isomerization and now by enzyme-catalysed isomerization. It can also be obtained from inulin, a polyfructose storage material used by some plants. Cereals contain small amounts, and so do tubers such as *Helianthus*. Commercial inulin is available from cereals, though the most promising source is the Jerusalem artichoke, or chicory in temperate climates and the yeast *Kluyveromyces marxianus* produces a specific $(2\rightarrow 1)$ inulinase [$(2\rightarrow 1)$-β-D-fructanfructohydrolase, EC 3.2.1.7]. (Fructans may contain either $(2\rightarrow 1)$ or $(2\rightarrow 6)$ linkages and vary much like starches.) It is possible that fructose production based on inulin with some optimization of the associated enzymes could provide a cheap source. Inulin tends to break down in the plant after harvesting, and in practice the syrup contains up to 20% glucose. It has already been used for specialized diabetic diets. The huge tonnages of starch available, and its by-product status in some situations make it unlikely that such processes will be developed on any significant scale in the foreseeable future.

Starch Degradation

DISRUPTION OF STARCH GRAINS

Starch grains vary in size and the proportions of amylose and amylopectin between species. The principal sources of starch are wheat, maize, potato and barley. Cereal starches contain 20–30% amylose, while potato starch has 26% amylose, the remainder being amylopectin (see Fig. 2.4). Varieties of maize are known containing (waxy) 99% amylopectin or (amylo) 25% amylopectin.

In the USA, the plant *Zea mais* is called 'corn'. In the rest of the English-speaking world, it is called 'maize', 'sweet corn' or rarely, 'Indian corn'. To confuse matters further, wheat is called corn everywhere except the USA. Glucose syrups based on maize starch are called 'corn syrups' in the USA and since they are mainly produced in that country we will use this name, bearing in mind that they have nothing to do with wheat. With this exception, the term 'corn' will not be used because of its ambiguity.

Starch production itself is highly developed around wet milling, which yields an exceptionally clean starch grain, needing little further handling before degradation.

The first step in starch processing is gelatinization. This is necessary because the usual enzymes used for starch degradation are unable to attack the intact grains. There seems to be one exception to this, the amylase from *B. licheniformis*. Patents have been issued which claim that it can be used at 60–75 °C without prior gelatinization. The reason for the success of this enzyme where other amylases fail is not known. The whole field of the attack of enzymes on heterogeneous gelatinized substrates has been neglected. Recently, some novel fluorescence methods based on fluorescence recovery after photo-bleaching have been applied to the problem of how amylase attacks starch. The technique offers the chance of progress in this and the similar problems of how proteases interact with protein gels and such things as protein bodies in fermentations (see Chapter 3).

High-amylose starches need higher temperatures for gelatinization, which begins at 50 °C for potato starch, and goes up to 68 °C for rice, sorghum and high-amylose maize. In this process, amylose forms the gel network, and the loss of amylose from the crystalline, structured starch grain is known as 'retrogradation'. The structures involved are hydrogen bonded, and although all starch grains contain small amounts of protein, it plays little part. The gels obtained by heating starch grains are irregular and contain swollen grains and fragments held in an amylose network.

It is possible to fractionate amylose and amylopectin by the use of butanol, and some starch derivatives have been used as thickeners and stabilizers in foods (see Chapter 5). There is also a substantial chemical industry based on starch derivatives.

In conjunction with gelatinization, the starch receives a preliminary hydrolysis. In the past this was acid-catalysed, but a problem with the use of acids was that the glucose content of the hydrolysate was variable and low. Full hydrolysis could not be achieved without, at the same time, producing undesirable flavours and colours. For this reason, enzymes have been widely used. Bacterial α-amylases are notable in that they are able to withstand temperatures > 100 °C and can function for long periods at 80–90 °C. They are thus ideal for use during the gelatinization step. It is known that the thermal stability of proteins is increased in the presence of polyols, but the effect is too small to account for these amylases. The bacteria of

Sweeteners

origin are not thermophiles and these enzymes probably have to be seen as one end of the wide distribution of thermal stabilities of naturally occurring proteins that are now known to exist. The liquefied starch is then further digested with an amyloglucosidase, to complete the hydrolysis as far as possible. It has been claimed that a *B. subtilis* amylase can completely hydrolyse a 20% starch suspension to glucose. Many glucosidases have been tried in sequence and in combination.

The product of hydrolysis is a turbid liquid, which is filtered to remove insoluble protein and lipids present as traces. Glucose monohydrate is made by crystallization of the concentrate. Because of the higher pH (6.0), and the milder conditions, the cost of the enzyme is more than outweighed by higher glucose yields and lower flavour and colour levels. The supernatant from final crystallization is used as growth medium for antibiotic manufacture. It would certainly be expensive to dispose of otherwise. Table 2.4 and Fig. 2.1 list some of the enzymes used in starch processing. By suitable choice it is possible to make hydrolysates containing maltose or maltotriose at high levels, rather than glucose. Pullulanase is sometimes used as a de-branching enzyme where the amylases cannot break down amylopectin. Because the bacterial amylases are relatively cheap, and also because of the high temperatures used, they have been used on a batch basis. There has been little incentive to develop immobilized enzyme reactors. They are feasible, though the interaction with the polymeric substrate is slower, and in the case of a β-amylase may actually alter the detailed specificity of the enzyme, to change the balance of the various glucose polymers present.

Glucose from Other Sources

The end product of the digestion of cellulose is glucose, and it is possible to make glucose syrups from this source. Since cellulose is the most plentiful organic material on earth, and is often prominent in wastes, it is an interesting question why it is not more widely used. One reason is that it occurs intimately mixed with lignins, which make it difficult to digest with cellulases. However, even pure cellulose is not easily broken down. Waste paper is the nearest approximation to a pure cellulose source and is likely to be the focus of commercial attempts. The problem is a dual one of our relative ignorance of how to design enzymes to act on solid or structured substrates (much the same problems as in starch or protein gels), and the lack of a suitable selection of cellulases. Since starches are able to supply all the glucose that is required at present, there is little commercial pressure to develop the biotechnology of cellulases (a mixture of endo- and exo-β-$(1\rightarrow 4)$-glucanases is believed to be needed) and lignases but when they do become available they could transform tropical agriculture. Some work on the degradation of lignins by peroxidases has been done.

Glucose Isomerization

Fructose is sweeter than glucose and although the difference is not large it is crucial

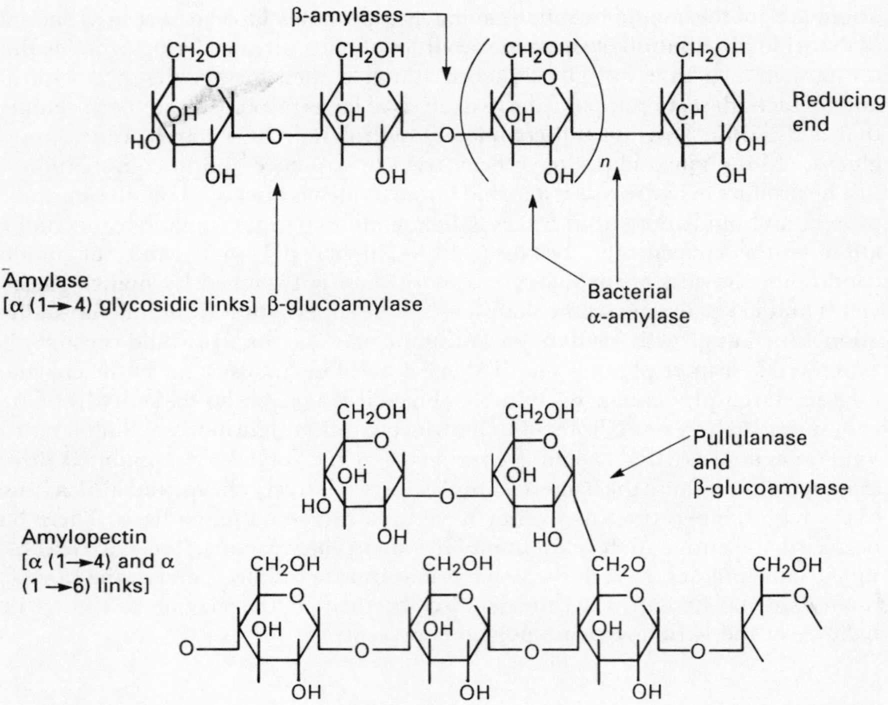

Fig. 2.1 The components of starch and the points of attack of starch degrading enzymes.

to applications in the confectionery field. Also, fructose has a higher solubility so that it can be transported and used as syrups with a higher solids content than glucose. High-fructose corn syrups, based on the use of corn (maize) starch have found large markets especially in the USA. One estimate is that their development led to an annual reduction in sucrose imports of $1 billion. Commercially, they are in direct competition in the sense that fructose syrups have displaced sucrose from applications. As the supply of a new ingredient develops it is usual to find it with small applications where it is much better than the old ones, and fructose is no exception. In Europe sucrose prices have been set in such a way that the use of high-fructose syrups is much smaller than elsewhere. More recently, processes for very high fructose levels have been developed, and while the price is higher, it has found uses in soft drinks. Other starches are cheaper than corn in some parts of the world, and wheat starch may be used for fructose manufacture in Asia and Australia, though inulin as a source may yet prove to be the best in remote areas. Table 2.5 shows the consumption of sucrose and syrups in developed countries, and charts the increase in fructose consumption.

The mammalian glucose isomerase system works through phosphorylated

Table 2.5 Consumption of sucrose and syrups in developed countries. (From C. Bucke (1979) in *Developments in Sweeteners*, **1**.)

Country	Sweetener	Consumption of calorie sweeteners in principal markets (million tonnes raw or dry basis)			
		1970	1975	1980	1985
USA	Sucrose	10.13	9.30	9.98	10.44
	HFGS	–	0.45	1.80	2.40
	Glucose syrup + glucose (crystalline)	1.73	2.30	2.13	2.16
Canada	Sucrose	1.05	1.05	1.16	1.21
	HFGS	–	–	0.05	0.12
	Glucose syrup + glucose (crystalline)	0.18	0.20	0.21	0.24
EEC	Sucrose	10.50	10.00	10.96	11.47
	HFGS	–	–	0.30	0.40
	Glucose syrup + glucose (crystalline)	0.80	1.10	1.14	1.24
Japan	Sucrose	3.00	2.80	3.04	3.33
	HFGS	–	0.05	0.33	0.45
	Glucose syrup + glucose (crystalline)	0.80	1.10	1.14	1.24
Total	Sucrose	24.68	23.15	25.14	26.44
	HFGS	–	0.50	2.48	3.37
	Glucose syrup + glucose (crystalline)	3.06	4.15	3.99	4.21

intermediates and requires ATP. It could not be used in large-scale processing. Bacterial xylose isomerases do not require ATP or any cofactor, and can be made to act as glucose–fructose isomerases by adding essential metal ions, either magnesium or cobalt. This discovery happened well before the techniques of gene manipulation were available. The enzyme (D-xylose ketol-isomerase, EC 5.3.1.5) from a variety of bacterial sources is expensive, and is used in an immobilized form. Much work has been done on optimizing its production. Xylose and xylans are both used to induce the enzyme. Because these are expensive sugars, organisms have been sought that would make the enzyme on a glucose substrate, both by deliberate mutation and otherwise. It is made in good yield by *Arthrobacter* and *Actinoplanes*, on glucose as the principal carbohydrate.

It catalyses the formation of the equilibrium mixture, of 55% fructose and 45% glucose. It does not act on maltose or higher polymers, and the final fructose level depends on the content of the feedstock. Current practice is therefore to obtain the highest possible level of glucose in the saccharification step. A chromatographic method for separating fructose and glucose is now in use based on cation-exchange columns which selectively retard fructose. It uses strong cation-exchange resins, in the calcium form, and is based on the fact that fructose can form a complex with calcium ions. An alternative, non-column method uses the complex in a different way. If sufficient calcium chloride is added to the mixture, the fructose complex crystallizes out first. However, calcium was then separated by electrodialysis, which is not really suitable for very large-scale use. The glucose is recycled until the residual oligosaccharide content gets too high, when it has to be discarded.

Mixing enables a range of products containing between 40 and 90% fructose to be made.

Immobilized Enzyme Reactors

Glucose isomerase is the only example of an immobilized enzyme in full-scale use in the food industry, though it will soon be joined by one using lipase (see Chapter 5). It is therefore the obvious choice for a discussion of this method of processing.

Immobilized enzymes are generally held to have advantages. They can be used in continuous processes. These can either be stirred-vessel or more often packed-column based. In the latter, the enzyme on an insoluble support is packed into a column and the substrate or a solution of it is pumped through. Continuous processes are more efficient than batch for large-scale processes. Moreover, while the apparatus is more complicated and therefore expensive, it can be much smaller. In the case under consideration the column can operate with a volume about one-fiftieth of that needed for a batch process of equivalent throughput.

Because it is smaller, it is easier to keep sterile and generally to meet food standards than with large open vessels. Also, providing the enzyme is stable enough it can be used repeatedly and more efficiently. Enzymes can be recovered from batch processes—some very ingenious methods based on magnetic particles have been developed—but the labour costs are so much higher that they are not used.

In the UK, glucose isomerase from *Streptomyces olivaceus* is acceptable, while in the USA and other countries *B. coagulans*, *S. olivochromogenes*, *S. rubiginosus* and *Actinoplanes missouriensis* are also used. They all have similar properties, needing a pH of 7–8 and temperatures of the order of 60 °C. Figure 2.2 shows the temperature dependence of the half-life of the enzyme, and the loss at higher temperatures is not compensated for by slightly higher reaction rates. It is important to have a low level of Mg^{2+} present since this improves the stability.

The packed-column enzyme has a half-life of between 70 and 100 days in use, and it is usual to run the column until its activity has dropped to about one-quarter of the initial level. The substrate is run in at about 50% solids, and has a residence time of only a few minutes in the column. In practice, the flow properties of the support material are crucial, and not easy to achieve with highly viscous substrates. The details are usually treated as commercial secrets and little has been published. It is usual to operate a number of columns in parallel, replacing them with fresh columns on a planned basis. The pumping rate, the enzyme activity and the column size are interrelated and numerous theoretical studies have been performed. In this process, since the conversion rate drops sharply as equilibrium

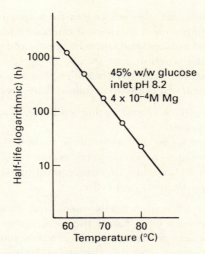

Fig. 2.2 The variation of half-life with temperature for an immobilized glucose isomerase. (From H.J. Peppler and G. Reed (1987), in *Biotechnology*, vol. 7a, Eds J. Rehm and G. Reed. Weinheim, VCH.)

is approached, a level of 42% fructose in the effluent has been selected as optimum rather than the equilibrium value of 55%.

The immobilized enzyme is very rarely a purified one, attached by covalent links to a suitable matrix. These are to be found in diagnostic reagents rather than the food industry. It is more likely to be a crude preparation of the organism with the enzyme attached in an unknown way to the cell wall. Many organisms that secrete enzymes in fact retain them on the outer side of the cell wall. Glucose isomerase is not a secreted enzyme, but seems to stick to some insoluble component. Again the details of immobilization tend to be commercial secrets of the suppliers.

There is a huge literature on enzyme immobilization and it is clear that most enzymes can be immobilized on a variety of supports by a variety of methods without much loss of activity. If we define immobilization to mean that the enzyme is not extracted from the matrix by the normal reaction mixture, and the activity as that shown in some standard assay, which may be under rather different conditions from those actually used in the process, then enzymes are remarkably easy to immobilize.

In commercial practice, immobilization tends to be carried out by the producers of the enzyme. This is because the simplest method is simply to disrupt the microorganism. In the original method for glucose isomerase, the organism, a *Streptomyces*, was heated in aqueous suspension to 80 °C, and the aggregated clumped cells used in fixed-bed reactors. A development of this was to mix the cells with polyelectrolytes, and suitable phosphates, such as calcium to make a harder floc.

A tendency to leach out the enzyme was stopped by treatment with glutaraldehyde. This bifunctional aldehyde is often used for food materials—it is used in making synthetic sausage skins from collagen—and is supposed to work by some sort of cross-linking reaction.

The following have all been used to immobilize glucose isomerase, essentially by adding a crude cell extract to the matrix followed by precipitation and drying:

- Anion-exchange celluloses, and cation exchangers saturated with Mg^{2+}, based on all the common resins
- Films of zein and collagen, and hydrophilic carbohydrates such as alginates
- Glass beads, magnesium carbonate particles, porous glass and other ceramic materials, sintered metal oxides including titanium oxide spheres, activated charcoal and clays, alumina and celite

The precise details of what is actually used, and how it is made remain unpublished. To emphasize once more, the most important property of immobilized enzyme materials, given that they have a reasonable activity level, is the flow properties. The current practice, almost certainly similar to the original clumped, heat-disrupted cells, does include the use of distributed small spheres to avoid channelling. It may also include the idea of replenishing the column by addition of fresh, soluble enzyme for adsorption which would obviously reduce downtime.

LOSS OF ACTIVITY

While there is no doubt that all enzymes used in flow reactors steadily lose activity, the reasons why they should do so are surprisingly obscure.

The definition of stability in the reactor context is an operational one, and the half-life under carefully defined conditions is universally used. But it is important to realize that it has only a tenuous connection with the thermodynamic stability discussed in Chapter 3. While it is likely that an enzyme with a low T_m (mid point temperature) of unfolding will also have a short half-life, this is not necessarily the case.

There are publications in which the rate of loss of activity has been used to assess the stability of enzymes, and conclusions drawn which should strictly only be derived from the thermodynamic stability. In more than one case, activity disappeared because of the unrecognized presence of proteases, and the rate has no fundamental significance whatever. This may well be the reason for some activity loss in reactors. Crude preparations in particular will contain traces of proteases, either from the enzyme preparation itself or from bacteria.

Another major reason is almost certainly 'poisoning', that is to say the presence of enzyme inhibitors in the feedstock. It is common to find that the life of the reactor is dependent on the precise feedstock used. In some instances the use of a 'pre-column' containing some cheap protein such as casein, will remove the inhibitors. Although it would be feasible to isolate the enzyme after it has lost its activity to see what chemical changes have taken place, this has never been done, and the nature of the inhibitors remains a matter of conjecture.

Another reason for activity loss is chemical change to the enzyme itself. Amide groups and thiol groups are the most labile, and especially at higher temperatures spontaneous hydrolysis or oxidation may occur. A well-documented case is lysozyme where the loss of a single amide group is a reason for irreversible thermal loss of activity. A comparison of amylases from *B. amyloliquefaciens*, and the thermophile *B. stearothermophilus* showed that in the first case, on heating in solution the enzyme unfolded, followed by the usual aggregation process. However, in the presence of substrate, which always stabilizes the enzyme, loss of amide groups at 90 °C was a major reason for activity loss. The thermophile showed similar effects, but including cysteine oxidation. The two enzymes have not yet been compared in reactor situations.

There is no good predictor of stability which could be used to test enzymes for their behaviour in reactors. Probably T_m is as good an indicator as any, but there are so many ways that activity can be lost that it is difficult to see how empirical testing in pilot-scale columns can be avoided. Even these have only limited predictive value for stability in large-scale installations. Enzyme reactors are peculiarly at risk from feedstock variation and in some applications it has proved necessary to apply strict quality controls, including enzyme stability assays in the presence of raw materials, all of which adds to the complexity of the process.

The Sweet Proteins

MONELLINS AND THAUMATINS

European explorers on the West coast of Africa have been reporting the existence of a 'miracle berry' for about 200 years. Well known to the locals, it was rediscovered in the 1960s as possessing remarkable flavour-modifying properties. This was the first time that the food industry had become aware of this berry, and it coincided with the banning of cyclamates. Cyclamates together with saccharin had been developed as purely synthetic sweeteners, which in combination could give a sucrose-like profile, particularly in soft drinks, at an acceptable cost. The banning of cyclamates in many countries was based on limited evidence and was controversial. Indeed, it is a good illustration of a basic principle, well known to food processors, that the decision to ban is a political one, taken by legislators who do not always base their views solely on test data. The cyclamate ban raised the question of where suitable sucrose substitutes could be found, which would be acceptable, and also must have made people wonder for how much longer they would be able to use saccharin.

The effect was to engender considerable interest in the 'miracle berry', which might well otherwise have been regarded as an interesting curiosity. Chewing the berries causes no particular taste sensation, but makes sour things taste sweet. Vinegar tastes like port, lemon juice like very sweet lemon drinks, and so on. The effect is persistent and can last for 6 h, changing the taste of most foods consumed in that period. It has no effect on salt or bitter flavours.

The miracle fruit shrub, *Synsepalum dulcifica*, grows widely in tropical West Africa, though it is not particularly common. It was also found to be growing in a

Florida botanical garden. The pulp in the berry contains the active material, miraculin, which proved difficult to extract. It is a glycoprotein of about 42 000 molecular weight and is the largest molecule known to affect taste. Small plantations were set up to produce enough of the protein for evaluation but after trials all products of this berry were denied food-additive status by the authorities. It remains of much interest to physiologists, since it may function by actually modifying the structure of the taste receptor proteins.

At about the same time it became widely known that the berries of two other West African plants, *Dioscoreophyllum cumminsii* and *Thaumatococcus danielli*, were extremely sweet. This is clearly a different phenomenon from the miracle fruit, but turned out to be due to some proteins present in the pulp, in the former case called monellin, and in the latter from the arils, called thaumatin.

Dioscoreophyllum cumminsii has proved to be very difficult to grow outside equatorial Africa, and although it is now possible to extract about 5 g of the active protein from 1 kg of fruit, other drawbacks quickly became apparent. The protein is not heat-stable, and was also unstable in cola-type drinks, probably because of the low pH of this kind of product. Even attempts at greenhouse cultivation ran into the problem that only male plants appeared!

Attention was therefore turned to *Thaumatococcus*, and its protein thaumatin. Fairly substantial plantations now exist as far apart as Togo and Malaysia and have provided enough material for extensive testing in numerous potential products. Moreover, the all-important food-use permission has been gained, and the first products have been launched in Japan. Thus thaumatin appears to be a class of sweetener with a future. We have already used thaumatin as one of the examples of gene transfer in Chapter 1. Apart from plans to produce it in yeast as a result of gene transfer, we should also consider whether there are possibilities of further improvement by, for example site-directed mutagenesis. It is also possible that monellin and perhaps other, as yet undiscovered, sweet proteins may become available by the same route.

Structure of Monellins and Thaumatins
Both monellin and thaumatin occur as a number of closely related forms, and are members of multigenic families. Although the relative amounts of the different thaumatins differ from one plant to another, a single berry has been shown to contain all five thaumatins. It is thus not a simple population polymorphism. The function of these proteins in the plant is unknown, but this kind of variability has been described for storage proteins.

The sequence of thaumatin is shown in Fig. 2.3, and compared with that of monellin. Monellin has two chains, held together by non-covalent forces. They are all very basic proteins, with pI about 11.5, and thaumatin is notable for the high cysteine content and the eight intra-chain disulphide bonds. There are five differences in the chain between thaumatins 1 and 2. The structure of thaumatin is indicated in Fig. 2.4. This is deduced from the sequence. Statistical analysis of known protein structures shows that certain amino acid sequences have a high probability of occurring in either helix, β-sheet, disordered structures, or β-turns. An alternative approach is to calculate directly the most stable structure of a

Sweeteners

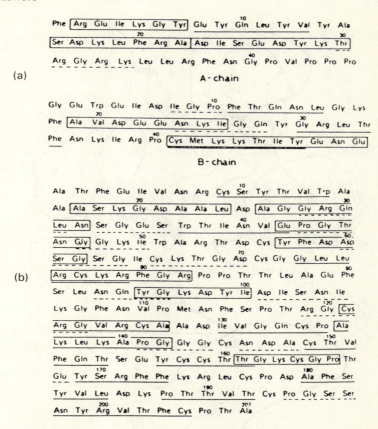

Fig. 2.3 The sequences of (a) monellin and (b) thaumatin A and B chains. The boxes surround sequences expected to be antigenic. (From H. van der Wel (1986), in *Developments in Food Proteins*, Vol. 4, Ed. B. Hudson. Barking, Elsevier Applied Science Publishers.)

particular sequence. Low-resolution X-ray data is consistent with these structures, and suggests that in the crystal, thaumatin occurs as a dimer form.

The most obvious question to ask about these structures, which at first glance show no obvious similarities, is what is it that leads to the sweet taste? The taste effect is dependent on the native structure. Heat disruption leads to loss of it. In thaumatin, fission of a single disulphide causes enough structural alteration for the same loss to occur. On the other hand, the two proteins show immunological cross-reaction, i.e. they have antigenic sites in common. It is believed that this involves the region of residues 25–31 in monellin B chain and 92–97 in thaumatin, both of which have a β-turn, and are typical antigenic sites. Since both proteins are also sweet, it is reasonable to suppose that these two sequences are also at least part of the active site for sweetness.

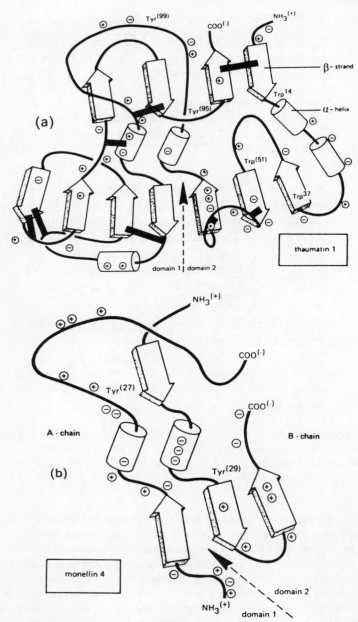

Fig. 2.4 Predicted structures for monellin and thaumatin. The disulphide links, β-sheets (arrows) and α-helices (cylinders) are all shown. (From H. van der Wel (1986), in *Developments in Food Proteins*, Vol. 4, Ed. B. Hudson. Barking, Elsevier Applied Science Publishers.)

Theories attempting to relate sweetness to chemical structure have been proposed in the past for a series of modified mono- and disaccharides and were generally accepted until the sweet proteins were discovered. The question is now regarded as open again, and some insight has been gained from selective modification of the residues of thaumatin. Methylation of thaumatin lysine residues, even up to 6 out of the 11 present had no effect on sweetness, and a similar result was obtained with monellin. The net charge and pI are drastically altered by this treatment. On the other hand, acetylation of the lysines did produce a drop in sweetness intensity, though no significant change in conformation could be detected by spectroscopic analysis. Succinylation, which replaces the positive charge on the lysine by a negative carboxyl one, had a large effect. A single substitution decreased the sweetness by half.

Many other similar substitutions of the basic groups were performed, with inconclusive results, and it is still not clear whether some lysines are essential or not. Amidation of carboxyl groups actually increased the intensity, in one case to 12 000 times that of sucrose. Iodination of some of the tyrosine had no effect, but insertion of iodine into all three resulted in loss of conformation. Modification of the single tryptophan led to conformation loss as well. Perhaps the most remarkable observation was that splitting a disulphide with cysteine led to the appearance of protease activity. Recently it has been shown that this is actually due to thiol activation of some contaminant proteases present in the preparation. Thus although it is known that the sweet proteins interact with high specificity with the taste bud membrane, there is no real indication of which residues are involved in the interaction. More precise information is likely to come from directed alteration of the sequence. Experience with proteases, for example, would suggest that the strength of the interaction will depend on two or three residues, probably well apart in the sequence.

Stability in Use
Since the sweet taste depends on the structural integrity of the protein, it might be expected that stability of the sweetness would present problems. In fact, thaumatin is stable indefinitely when freeze-dried and in aqueous solution between pH 2 and 10 providing bacteria are excluded. The solution can be pasteurized, and is most stable to heat between pH 3 and pH 6. It is possible that in some cases heat disruption is reversible on cooling, for thaumatin, but this would depend very much on the precise environment. A complex with alginates is said to have a better heat stability. There is little to choose between thaumatin and monellin in heat stability, and the eight intra-chain disulphides of thaumatin do not appear to confer exceptional stability. They may however make re-folding easier.

It is difficult at present to point to any features which might be the subject of modifications. Although the gene has been cloned and expressed in both bacteria (*E. coli*) and yeast, it is not likely that modified versions will show sufficient advantage to justify replacing thaumatin in the near future. Even if a superior form is discovered it will take several years to obtain food-use permission.

So far, the only firmly established use for thaumatin is in Japanese chewing gum, supplied from plantations in Malaysia. It has synergistic flavour-enhancing effects on peppermint flavours. One of its drawbacks is a rather slow perception time. In other words, the flavour can develop slowly, and this is something of a difficulty in soft drinks. Combination with other sweeteners and reformulation can overcome this, and as supplies increase thaumatin will almost certainly appear widely in low-calorie applications.

Aspartame

Aspartame is the name given to a dipeptide of aspartic acid and phenylalanine methyl ester (Fig. 2.5) and is remarkable, for a small molecule, in the number of aspects of biotechnology which may become involved with its manufacture.

It has been known for a long time that amino acids and peptides can be sweet, and aspartame was noted in the 1960s as possessing exceptional intensity. It was first produced by a chemical synthesis, as shown in Fig. 2.5, by reaction of L-aspartic acid anhydride with L-phenylalanine methyl ester. Both the α- and β-dipeptides form and must be fractionated. Much work, and hundreds of patents on detailed improvement of this process by using selected solvents and temperatures to improve the yield of the α-dipeptide, have led to the current chemical route.

There was a major battle over regulatory approval, which took more than seven years to achieve. Biotechnology enters because even the chemical synthesis needs L-aspartic acid and L-phenylalanine as starting materials.

A major advantage of a fermentation process is of course that it can be made to produce exclusively L-isomers, thus avoiding one of the main difficulties of chemical synthesis, which tends to form D,L mixtures. Established large-scale processes using *trans*-cinnamic acid and phenylalanine ammonia lyase supply aspartame manufacture. Both large-scale fermentation in stirred vessels and an immobilized form of the yeast enzyme are in use. A variety of yeast containing high levels of the enzyme has been selected for. However, because of competition and the fact that aspartame patents will shortly come to an end, we have an interesting position. It is common enough for novel biotechnology processes to be in competition with old established non-biotechnology methods, but rivalry between different fermentation processes to make the same product is unusual. Recently, a *Corynebacterium* has been found that can make phenylalanine from phenylpyruvate, formic acid and ammonia, via acetamidocinnamic acid. An outline of the metabolic pathways is given in Fig. 2.5. What makes this significant however is that the gene for the ACA acylase II has been cloned and sequenced, and a promoter found which leads to hyper-expression in a mutant of *Corynebacterium*, which was able to make L-phenylalanine very efficiently.

The motive for developing alternative processes is often to avoid patents, and since this is an area heavily covered by patents, which distort competition based purely on cost criteria, it may be a long time before we know which process is the most effective. L-Aspartate can also be made by fermentation, from fumaric acid. This is done typically with vermiculite-immobilized *E. coli* strains with high aspartase activity and has been the main production method in Japan since 1973.

Proteases can be used to make the peptide link, and oddly enough can reduce the advantage of fermentative manufacture of L-amino acids. This is because

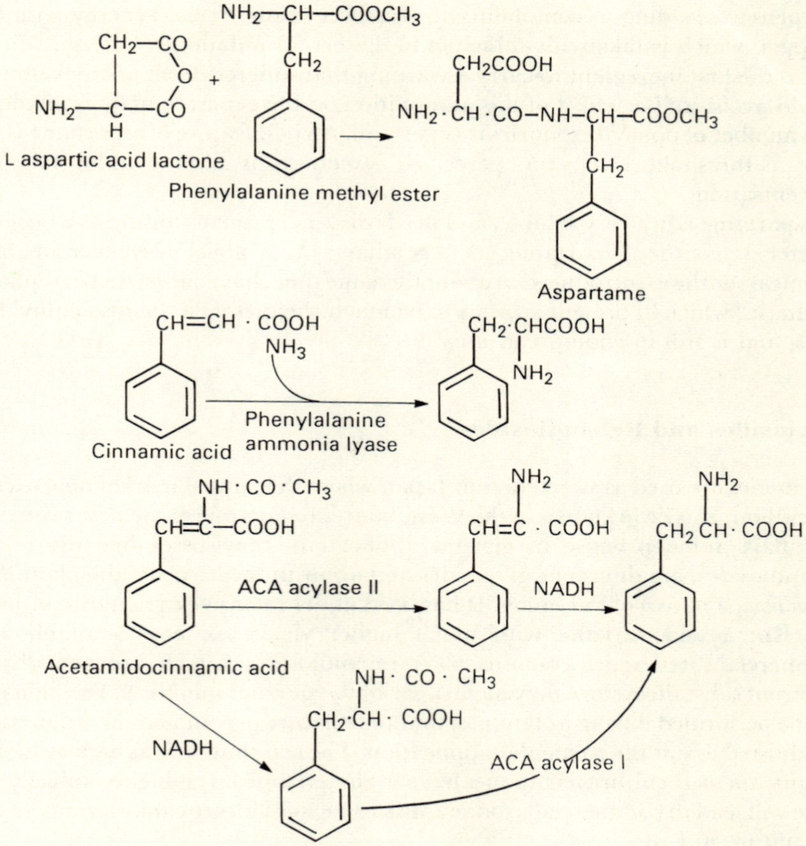

Sweeteners

Fig. 2.5 Routes to aspartame. (Top) chemical synthesis; (bottom) two enzyme-based routes to L-phenylalanine.

proteases can selectively link the L-form in a D,L mixture. Reaction of high concentrations of N-acylated aspartic acid with the methylester of phenylalanine in the presence of thermolysin leads to an insoluble product, helping to reverse the normal course of proteolysis. Perhaps the most interesting proposal for the biotechnology of aspartame is to make an organism produce a copolymer of aspartyl phenylalanine, presumably by insertion of a GAUUUU sequence into a suitable vector. Following isolation and fission with an enzyme such as chymotrypsin the remaining methylation should be feasible. This is not yet in use as a production method, and may never be.

Aspartame in application is said to be the most sucrose-like of the new sweeteners in its detailed taste profile. Its main drawback is its acid instability which makes applications in the high-use field of soft drinks problematical. Because it is slowly lost, there has been a tendency to put in rather too much to compensate, leading to complaints of over-sweetening. It has synergy with fruit flavours, which is taken advantage of in dessert formulations.

It is the first ingredient to carry a warning that sufferers from phenylketonuria should avoid it. The effect of this admonition on the general public is unknown. The number of phenylketonurics is very small. An odd feature of aspartame is that over a threshold level, the perceived sweetness is almost independent of concentration.

Aspartame is not very soluble, and needs dispersing agents for use as a beverage sweetener. Neither aspartame nor Acesulfame K, a novel sweetener made by chemical synthesis introduced at about the same time, have succeeded in replacing saccharin, which at present appears to be much cheaper on a 'sucrose-equivalent' basis, and is still in widespread use.

Steviosides and Rebaudiosides

Steviosides are used as sweeteners in Japan where the annual market now exceeds £5 million. It is little known in the West, but represents one of the new sweeteners that have some promise in general application. Steviosides like the related rebaudiosides, are diterpene glycosides and occur in the leaves of the plant *Stevia rebaudiana*, a native of Paraguay. It has been under intensive agriculture in Japan and Korea and varieties with much higher yields are now available. The commercial 'stevioside' contains 55% rebaudiosides, which are actually the preferred substances since they have superior flavour and solubility. Fractionation can be performed during isolation so that the mixture in commercial products can be adjusted to suit the particular application. The leaf content is as high as 12% by weight, and cell-culture techniques have been developed. It is highly unlikely that these will lead to commercial production, since tissue culture cannot compete with agriculture in cost.

It has quite good acid stability and can be used in soft drinks but its main use is in chewing gum. It also offers some advantages in sweetening fermented food such as soy sauce; unlike sucrose it does not get involved in the fermentation. The sweetness profile is said not to be good and it is likely to be used in combination with other materials. Enzymic hydrolysis of the glycosides with amylase has been used to convert stevioside to rubusoside. The latter can then be converted by chemical synthesis to a variety of other glycosides, especially rebaudioside A, which has most applications. Some of the structures involved are shown in Fig. 2.6. The background biochemistry of these glycosides is almost completely unknown, and it should not be difficult to adjust the pathways so that mainly rebaudioside A is produced. The Food and Drug Administration in the USA has refused permission for stevioside use in foods.

Conclusions

The branch of the food industry concerned with sweeteners illustrates perhaps

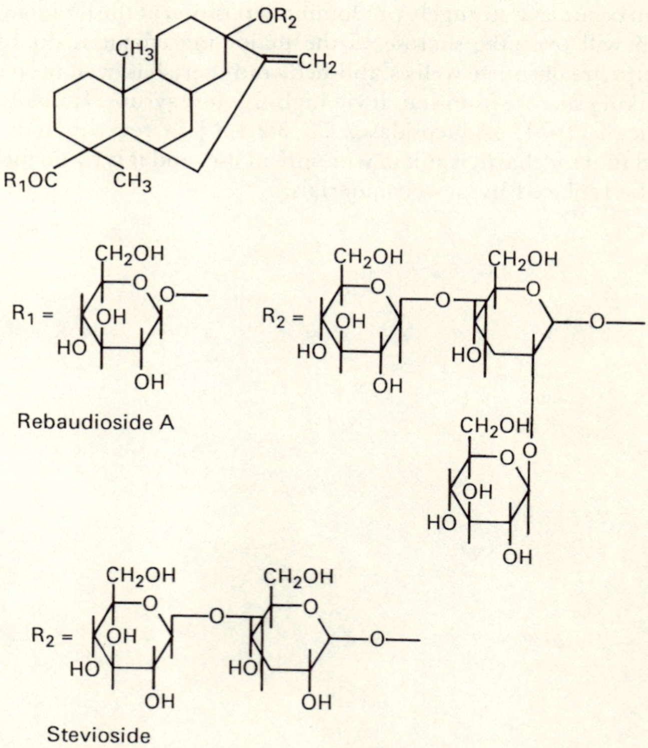

Sweeteners

Fig. 2.6 Structures of the triterpene alkaloids rebaudioside A and stevioside.

better than any other the extraordinary diversity of the ways in which biotechnology can affect just one ingredient of foods. The production of high-fructose corn syrups is the prime example of the application of an immobilized enzyme, just as the older methods for digestion of starch are the principal examples for the batch use of enzymes.

Aspartame brings in fermentation for the manufacture of L-amino acids, while the rebaudiosides introduce the complex problems of the biochemical pathways involved in synthesis of minor plant components. The sweet proteins exemplify the full-scale transfer of genetic material to a more congenial organism for production of the desired protein. All this is in competition with sweeteners made by chemical synthesis.

It is quite impossible to foresee the outcome, except that it seems inevitable that some sucrose will be replaced by novel compounds. We have not considered sucrose derivatives made by purely chemical means, or alcohols such as sorbitol or lactitol since there is no reason to suppose that biotechnology will be involved with them, except possibly as competing ingredients. What will probably happen is that each new sweetener will find its own niche, where it is clearly superior. (Japanese chewing gum occupies a strangely predominant position at the moment.) Whether any of them will overtake sucrose as the major ingredient is doubtful. High-fructose syrups are of course well established, and there has even been one project aimed at making sucrose from starch via high-fructose syrups. It involves running sucrase [glucan $(1\rightarrow 4)$-α-glucosidase, EC 3.2.1.3.] in reverse. It is also to be remembered that saccharin is still in widespread use, and it is by no means certain that it will be replaced by newer materials.

Chapter 3

Mainly Proteins: Proteases, Gels and Fermented Foods

Introduction: Food Texture

A lot of foods depend mainly on protein for their textural characteristics. Structures built up from proteins have been identified as the most important in dairy products, such as cheese and yoghurt, virtually all meats and processed meat articles, a great variety of fermented foods based on legumes and other beans and seeds. Proteins are also important, though perhaps less so than starch in cereal-based products like bread.

In gross composition, the seeds that form the staple foods for most of mankind fall into two broad categories: those containing protein, lipid and inert carbohydrate such as cellulose cell walls (sometimes called 'fibre'), and those containing, in addition, some starch. In the latter, starch may play a significant role in texture.

This is of course an over-simplification, but goes some way to account for the parallels between dairy products, made from milk in the West, and a range of soy-based foods in the Far East. They are both based on a mixture of lipid and protein in roughly equal proportions, with fermentable carbohydrate besides. The difficulty of defining the textural or structural properties of foods is borne out by the large number of more or less imprecise terms used. Mouthfeel, chewiness, tender or tough, textured, rubbery, fragile, structured, gel-like, are only a selection of the adjectives. One rather desperate attempt to quantify texture consisted of a set of false teeth fitted with a motor and pressure sensors. Most people have used more or less sophisticated rheological measurements to try and describe food texture. This has been successful, particularly in defining parameters for use in quality control during manufacturing operations. It has been less successful in attempting to relate food-texture properties (as defined in rheological terms) to consumer preference. Indeed one rather disheartening conclusion from extensive work is that the general public is much less discriminating about the texture of foods than it is about other attributes such as flavour or price. Nevertheless food manufacturers believe that food structure, defined in rheological terms, is important and it is a prominent theme in food research. But this approach does create a problem for the biotechnologist. Proteins and the actions of proteases on them are understood and studied mainly in molecular terms. There is a great gap between our knowledge of proteases and proteins and how they interact, and the kind of structures proteins form, and the rheological measurements used to describe foodstuffs in consumer and manufacturing terms. Despite much work, the gap has not been closed and the problem is in fact an extremely difficult one. We know only in the most general terms how to control the kind of structures that involve proteins by varying process conditions or by the use of proteases.

In this chapter we will be exploring how proteases might be used and what sort of properties they should have in order to improve and extend protease-dependent processes in the food industry. This understanding in its turn requires a knowledge of the basic structure-forming properties of proteins. Only by doing this can we hope to bring the resources of biotechnology to bear in making new or improved enzymes available.

Protein–Protein Interactions and Structure Formation

Peptide chains are rather good at interacting with other peptide chains and that includes other parts of the same peptide chain. The most important initial distinction is to differentiate between covalent and non-covalent interactions. There are a variety of both, of significance in foods, and some of the covalent ones are under enzymatic influence.

To deal with non-covalent ones first, these are influenced by the composition and the configuration of the peptide chains, and are the most likely to be affected by processing conditions.

NON-COVALENT INTERACTIONS

Hydrophobicity
Amino acid residues are characterized as hydrophobic or hydrophilic and quantitative estimates for each residue can be derived from studies of the solubility of amino acids in water. Table 3.1 gives some numerical estimates, but it is useful to classify the residues into two broad categories. All the charged side chains are hydrophilic, while all the paraffin side chains, such as leucine and valine, are hydrophobic. So are all the aromatic residues, methionine and cysteine. Note however that cysteine has a pK around 8.0 and when ionized will be hydrophilic. Serine, threonine and the amides are usually counted as hydrophilic. Thus it is possible by using the amino acid composition to assign a hydrophobicity index to a protein. It is a matter of observation that peptide chains in water tend to fold up in such a way that the hydrophobic residues are on the inside of the molecule while the surface is covered by the hydrophilic residues. Not only that, but in globular

Table 3.1 The hydrophobicity of amino acid residues*

Residue	Transfer free energy (kJ/mole residue)		
	Water to ethanol	Protein folding	Water to vapour
Ile	21	2.9	9.0
Phe	21	2.1	−3.2
Val	12.5	2.5	8.4
Leu	14.4	2.1	9.6
Trp	27.2	1.3	−24.7
Met	10.5	1.7	−6.3
Ala	4.2	1.3	8.0
Gly	0	1.3	10.1
Cys	0	3.8	−5.2
Tyr	18.8	−1.7	−25.6
Pro	6.3	−1.3	−
Thr	2.1	−0.8	−20.6
Ser	−2.1	−0.4	−21.3
His	4.2	−0.4	−43.3
Glu	−	−2.9	−43.0
Asn	−6.3	−2.1	−40.5
Gln	−4.2	−2.9	−39.4
Asp	−	−2.5	−45.8
Lys	−	−7.5	−40.0
Arg	−	−5.8	−83.6

*The three columns are estimates of the free energy change for transfer from water to ethanol representing a fairly hydrophobic solvent, for transfer from water to the interior hydrophobic core of a protein, and from water to vapour. (Data from F. Franks (1988) *Characterisation of Proteins*, Humana Press, NJ)

proteins, soluble in water or salt solutions, the composition is such that there are just enough hydrophilic residues to cover the surface. Peptide chains that do not have enough hydrophilic residues to cover the surface simply stick to the corresponding hydrophobic patches on another chain. This is how multi-subunit proteins arise. If the surface is still hydrophobic after such interactions then the protein will become insoluble in aqueous media.

Like all generalizations about protein structure it is dangerous to take it too far. In fact, where detailed structures are available from X-ray crystallography, it is clear that even soluble proteins have a mosaic structure on the surface with some hydrophobic patches. Also the occasional charged group can be found deep in the structure, and in at least one case, carboxypeptidase, a water-molecule is trapped inside. The idea of the hydrophobic 'bond' comes from work on the solubility of paraffins in water and for our purposes it is sufficient to note that they (that is hydrophobic residues) prefer to be in contact with each other, when surrounded by water, for entropic reasons. A feature of the interaction is that it becomes weaker when the temperature is decreased. It cannot exist in the absence of water.

Charge–Charge Interactions
Proteins carry both positive and negative charges, almost entirely on their surface, and the charge is of course dependent on the pH of the medium. The charge is a major factor in determining the solubility, and it appears to be the net charge that matters. Thus the solubility tends to be lowest at the point of maximum charge, but zero net charge, the isoelectric point. On the other hand, in the structures themselves, a statistical survey of known structures shows that about 70% of charged groups are so close to a charge of the opposite sign that they can be regarded as an ion pair. Charge–charge interactions, both repulsions and attractions, are undoubtedly important in subunit interactions, and in general account for the ionic strength sensitivity of protein-based structures.

Hydrogen Bonds
Hydrogen bonds are important in the helical and sheet structures found in globular proteins, as well as in special structures like the collagen helix. They can also be important in subunit contacts: they are responsible for the very strong interaction of insulin subunits, for example.

The exact configuration adopted by a protein is the result of the interplay of the three types of interaction described above, together with exclusion forces. The latter amount to allowing for the fact that one cannot put two residues in the same place. If we consider for the moment a protein with a definite structure such as an enzyme, the stability is the result of small differences between large quantities. The whole structure can easily be pushed over into a different one by a comparatively small change in the environment. This is why a small change in temperature can bring about a complete alteration of the configuration of a protein and since this may quite change the nature of the residues exposed to the solvent, may drastically alter the solubility. This is equivalent to saying that altering the configuration alters the way in which the protein chain will interact with other protein chains, and hence the formation of structures. The forces involved in chain–chain interactions are the same as those that maintain the configuration of the folded chain by interaction of different parts of the same chain.

COVALENT CROSS-LINKS

Although cross-links leading to gelation can be entirely non-covalent, in food systems there will also be covalent ones of various kinds. Their presence can easily be detected because solvents such as strong urea solutions, or sodium dodecyl sulphate, break all except covalent links. SDS gel electrophoresis is therefore well adapted to detecting their formation. They are important because they are individually, though not necessarily in aggregate, stronger than non-covalent bonds. They are rarely broken by heating and are therefore the dominant form in thermo-stable gels, which are so important in food processing.

Plasteins

Before the ribosomal mechanism of protein synthesis had been discovered, it was thought that it might be due to a reversal of proteolysis. To test this, concentrated mixtures of peptides, produced by acid hydrolysis, were treated with proteases. In fact the average molecular weight did then rise, but on further investigation it seems that most of this increase was due to linkage of glutamate side chains to lysine residues (Fig. 3.1). This is a peptide link. It needs few cross-links to cause a large increase in molecular weight, and the size of peptides formed was not very large. More conventional peptide chain links, on α-amino groups, can also form in a few special cases. The resulting mixture was called 'plastein'. This kind of cross-linking may well be involved in fermented soy foods, though no reports have yet appeared. There has been some interest in plasteins more recently, and it is possible that a more careful selection of enzymes might lead to a useful process. While proteases differ in their ability to form plastein, no optimization has been done.

Inter-Chain and Intra-Chain Disulphides

Free thiol groups on proteins are relatively uncommon, often highly reactive and, when present, involved in enzyme active centres. Their reactivity is conformation-dependent, and one of the results of unfolding is often to activate a thiol group. Most cysteine occurs in the oxidized form cystine, and thiol tends to undergo exchange reactions with the disulphides. Figure 3.2 shows some of the reactions that happen, and they lead to inter-chain links. This is probably the commonest cross-linking reaction, and since it can be a chain reaction it can produce gelation. The best known case of this is for serum albumin, which also exemplifies another common feature. When isolated it is found that the free thiol is partly cross-linked to glutathione. This tripeptide is present at high levels in all animal tissues, and has never had a clearly known function. Recently an enzyme called 'protein disulphide isomerase' has been characterized. Broadly, it catalyses disulphide-thiol interchanges, and is believed to be a part of the protein folding mechanism *in vivo*. It is known that proteins with three or four intra-chain disulphides can fold with 'wrong' cross-links which then slowly rearrange to give what is presumably the most stable form. The enzyme speeds up this process. It occurs in both animals and plants and clearly offers possibilities of controlling cross-linking, if it were available. It might also find application in hairdressing, since perming involves rearranging the disulphides of keratin, as well as in wool processing. The enzyme has been cloned and will doubtless be tried soon. It has already been investigated as an adjunct to bread-making (Chapter 4) but is not yet available in sufficient quantity for pilot-scale trials.

Cysteine chemistry is very complex and while disulphide exchange is probably the main reaction, reactions with molecular oxygen to give cysteic acid and with sulphite to yield S-sulphocysteic acid are also important. Sulphite is the most effective reagent for splitting disulphide links, while thiols such as mercapto-ethanol are poor and require very high concentrations to achieve what are in effect disulphide rearrangements.

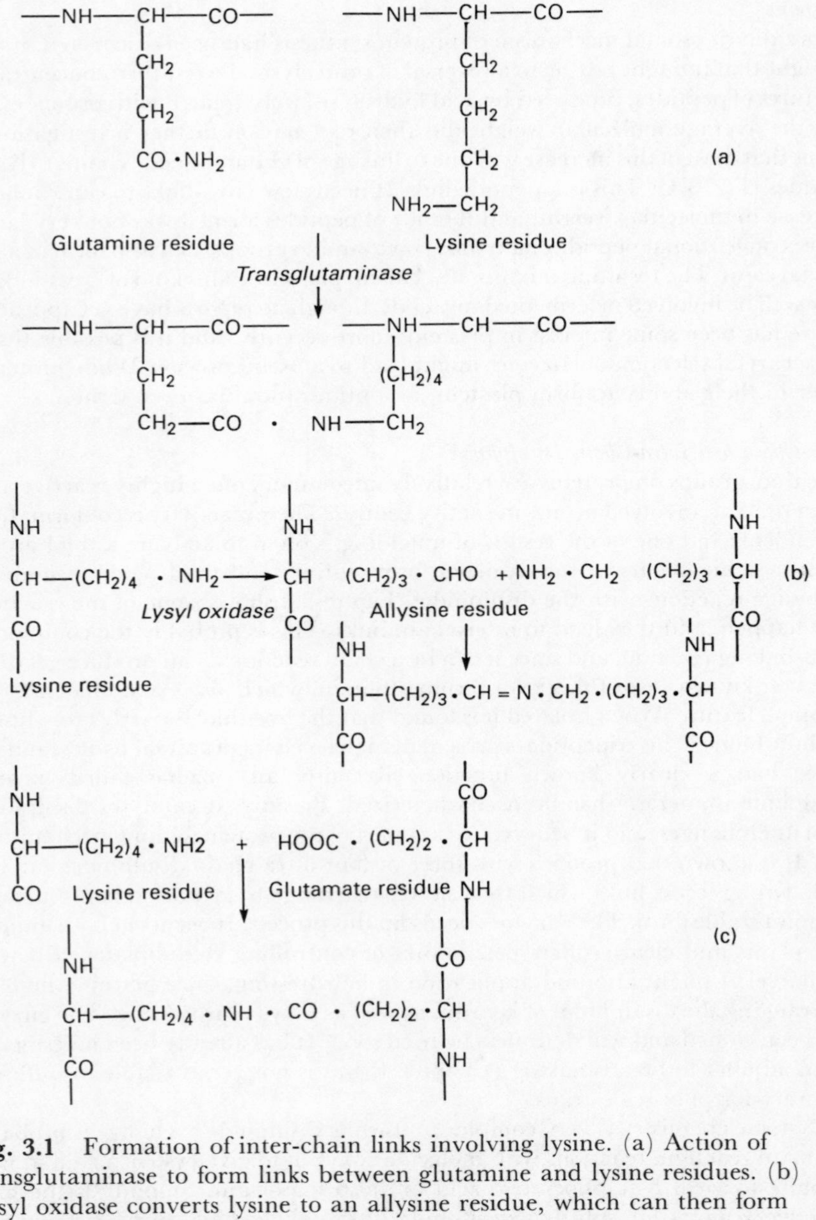

Fig. 3.1 Formation of inter-chain links involving lysine. (a) Action of transglutaminase to form links between glutamine and lysine residues. (b) Lysyl oxidase converts lysine to an allysine residue, which can then form cross-links with further lysine residues. (c) Links formed in the plastein reaction between glutamate and lysine.

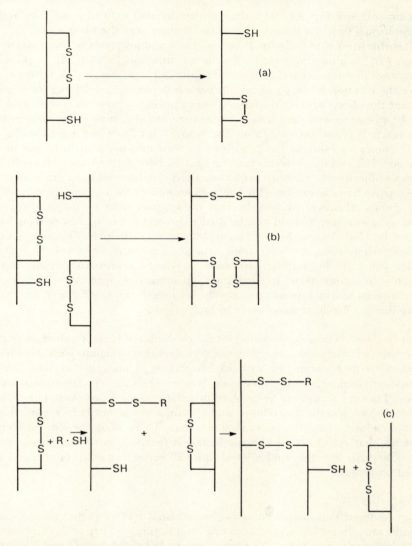

Fig. 3.2 Some disulphide rearrangements known to occur in unfolded proteins. (a) Internal rearrangement, (b) interchain link formation, common in seed proteins, and (c) thiol catalysed inter-chain disulphide formation, which can lead to gelation. Serum albumin is the best known example.

Selected Enzymes

Transglutaminase. This enzyme, which occurs widely in animal tissues, catalyses the formation of links between glutamine residues and lysine with the elimination of ammonia (see Fig. 3.1). It is thus potentially able to form cross-links, and its physiological function is to do this in the formation of the fibrin clot.

It is able to cross-link caseins, β-lactoglobulin and soy proteins, but is said not to react with ovalbumin, serum albumin or immunoglobulins. It is probably significant that those that do not react are all rigid highly structured proteins, and since the reaction is likely to be conformation-dependent, unfolding them might change this. Acetylation of the lysines, not surprisingly, prevents cross-linking but can be used to control the reaction, since unreacted glutamine in the same protein can react. It is also possible to use the α-amino group of free amino acids, and methionine, for example has been incorporated into soy protein by use of this reaction. It is said that proteases cannot split the bond formed, and this might be a suitable objective for modified protease activity. γ-Glutamyllysine can be ingested and replace lysine in rat diets, but the mechanism by which lysine is regenerated is not known. However, it may be that quite apart from its use as a texture-producing enzyme, it could also be used to protect lysine against cooking losses caused by Maillard reaction with sugars or reaction with lipids. This is yet another enzyme where more extensive trials depend on a better supply of the enzyme, which will probably require expression in yeast. Unfortunately it will not be possible to demonstrate its usefulness in industrial application before this expensive step, which would not ideally be undertaken until utility was firmly established. This dilemma is likely to be frequent.

Lysyl Oxidase. This enzyme catalyses the oxidation of lysine residues to give an aldehyde (allysine) which can then undergo a series of reactions with other lysine residues, with histidine, or with carbohydrates. This, together with similar compounds formed from hydroxylysine, create cross-links in collagen, and elastin, *in vivo*. They are thought to be involved in collagen-ageing processes. The enzyme appears to be specific for collagen and elastin, and has not been regarded as a candidate for controlled cross-linking processes. There would be some interest in enzymes that could reverse the condensation reaction, but there appear to be none. Nevertheless, this kind of cross-link will occur in meat products and gels based on gelatin.

Cross-Linkers

Other common food components can generate small molecules that are capable of forming cross-links. Lipoxygenase acting on unsaturated fatty acids can produce dialdehydes which cross-link lysine residues. In fact dialdehydes are deliberately added in processes for making sausage skins from collagen. The hydroperoxides also made, almost certainly become involved in disulphide exchange reactions. Polyphenol oxidase produces quinones from substances such as catechol, and other phenolics. Since these are present in plant tissues, but not animal, they are an extra hazard of plant raw materials. The precise nature of the reactions appear to be unknown, but lysine side chains and thiols are involved, and insolubility of plant protein preparations is often ascribed to cross-linking by polyphenolic materials. They are also believed to damage the nutritional value by making the protein immune to proteolysis as well as by selectively damaging lysine and cysteine.

Stability

Macromolecules in general, and proteins in particular, can be classified as either flexible chain or rigid and the rigid ones can be further subdivided into those whose shape can be described as ellipsoids of revolution, or as an assembly of beads. Proteins can be moved from one class to the other by varying the conditions, though some random ones will never be rigid, and some very small ones such as lysozyme and α-lactalbumin are only capable of forming one bead. Large ones like the fatty acid synthetase described in Chapter 1 do have a structure like a string of beads joined together by loops, and the exposed loops are susceptible to protease attack. So are flexible coils, but rigid globular proteins are very resistant. This is purely a conformational effect—unfolding makes them far more available. One of the best examples is α-lactalbumin, where carboxypeptidase can only remove one residue from the C-terminus in the folded or 'native' form, but at least seven when the chain is opened up. An endopeptidase such as trypsin will only attack about 2% of the bonds for specificity reasons, and in a typical domain of about 200 residues, the probability is that the four susceptible ones will not be accessible. Since protease attack is at the heart of structure formation in fermented foods it is essential to have some understanding of the protein conformations in the substrate. This is so both because it affects the susceptibility to the enzymes but also because the nature of the protein–protein interactions are determined by the conformation.

Proteolysis itself affects the structure but much less than might be expected. Multi-site enzymes like FAS can be chopped up into domains that still retain their activity, while small enzymes like ribonucleases can suffer two or three breaks without coming apart. In other words the continuity of the peptide chain is not especially important in maintaining a folded structure and limited fission will not generally bring about unfolding. It will be evident that in most fermentation processes the unfolding precedes the fission, and is brought about by heating or other destabilizing treatment.

Limited proteolysis may not bring about unfolding but it does affect the stability—it reduces it. Figure 3.3 shows a typical heat-induced unfolding curve, in this example for ovalbumin. The structure changes over a narrow temperature band, cooperatively, from a highly ordered structure to a flexible, more random one. If it is an enzyme the activity is totally lost as a result. The form more stable at the higher temperatures is sometimes described as a random coil. This is not strictly accurate, since the random coil is a precisely defined structure seen in simple homopolymers. It is described by a particular statistical distribution of the distance between the two ends of the chain, which can vary from contact to the full length. The random coil is a more or less spherical particle (the term is occasionally used, quite wrongly, to describe extended chain conformations) and so are unfolded globular proteins. It has been shown that peptide chains in solvents such as 6 M guanidinium hydrochloride do approximate to random coils, providing any intra-chain disulphides are broken.

80 Biotechnology in the Food Industry

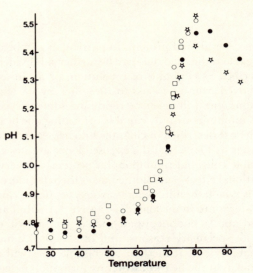

Fig. 3.3 Unfolding curves for four different samples of ovalbumin. In this example the pH change in unbuffered 0.1 M sodium chloride solution was used to follow the conformational change. The pH change is due to alteration of the pK of ionizing groups as a result of the change in their environment during unfolding. Numerous other properties such as spectra or hydrodynamic behaviour can also be used. In the case of enzymes, activity loss is also associated with unfolding, but cannot usually be measured over the whole range of temperatures, so that separate assays only measure irreversible loss of activity. The mid-point of the curves T_m, in this case is at about 71 °C and is used as an estimate of inherent stability.

THERMODYNAMIC STABILITY

The sort of measurements shown in Fig. 3.3 can be used to estimate an equilibrium constant for the process,

$$\text{folded} \rightleftharpoons \text{unfolded}$$

and there are many examples where the process is known to be fully reversible. In conjunction with calorimetric measurements this leads to an estimate of ΔG and the way in which it varies with temperature. Figure 3.4 shows some results for typical seed globulins, which are themselves typical of all globular proteins. They illustrate two very important points. First, ΔG is the numerically small outcome of the difference between the much larger ΔH and $T\Delta S$. Secondly, it clearly goes through a maximum, so that there is an optimum temperature for stability. The implication that proteins unfold at both high and low temperatures has been borne out by experiment. Chymotrypsinogen for example unfolds at about 60 °C and at −33 °C in supercooled water. The optimum tends to be at 0–40 °C, though there is no inherent reason in peptide structure why it should be in this range. The

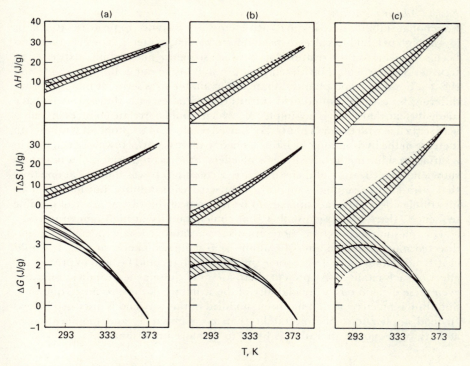

Fig. 3.4 Free energy (ΔG) enthalpy (ΔH) and entropy ($T\Delta S$) change associated with thermal unfolding of (a) soy bean, (b) *Vicia faba* and (c) sunflower storage proteins (protein body proteins). Note that ΔG represents a small difference between two much larger quantities ΔH and $T\Delta S$, and also note that it passes through a maximum. This implies that there is a temperature of maximum stability. (From V. Tolstoguzov, private communication.)

result is presumably the outcome of natural selection. Thermophiles for example, have a different stability pattern, with a broader range rather than a higher optimum.

It is important to note that stability estimates based on ΔG are for the thermodynamic stability, which one might suppose has some fundamental connection with the structure. We have already met the quite different functional stability definitions, based on half-life of activity in reactors, in Chapter 2. The two must not be confused, and there is not necessarily any connection or correlation between them.

The measurements shown in Fig. 3.3 can only be made in solutions that are so dilute that aggregation does not occur. In food systems which tend to be concentrated, the exposed hydrophobic patches of the random form will lead to aggregation and either coagulation and phase separation, or gelation. The former is the phenomenon called 'denaturation' in the past.

Factors Affecting Stability

Functional stability may involve such factors as resistance to protease attack, or low or high pH, and has already been referred to in Chapter 2. However there has been some work on the relationship between sequence and stability expressed in terms of ΔG, particularly with temperature stability in mind. Both lysozyme and the α-subunit of tryptophan synthase are available in a series of mutant forms, differing by a single amino acid substitution. Studies with these proteins have made it clear that there is no simple rule which would correlate changes in amino acid sequence with changes in ΔG. For buried residues, in tryptophan synthase, an increase in the hydrophobicity index tended to increase ΔG. However, a change in a surface residue might have little or no effect. Studies on thermophiles have also shown no single feature of their structure in common. It was thought at one time that intra-chain disulphides might be connected with stability, but this is not an invariable effect. The fact appears to be that out of the vast number of possible sequences, there is a large number that confer stability at high temperatures.

The question has become more important because it is now possible, by site-directed mutagenesis to alter the amino acid sequence almost at will. It would clearly be helpful if there were some simple rules that could be used to predict the effect of such changes. At present there are only quite general indications, and even these depend on the availability of the structure of the enzyme in question. Subtilisin is an enzyme that has been modified in this way and is discussed below. In food applications, thermostability can be a nuisance, and modification of activity by sequence alteration is likely to be more important.

Protein Gelation

The formation of gel like structures by proteins is a fundamental feature of a large number of food-manufacturing processes, particularly those that involve protein-based structures. While it is true to say that there remains much to discover about these gels, they represent a major area in which biotechnology might be applied. The examples of cheese and yoghurt given below are sufficient to show the way in which enzymes can be used to modify the structure-forming abilities of proteins. We must now examine the more general and fundamental aspects of protein gelation in order to explore the way in which a more structured, and certainly less empirical, attack on the problems might be made.

For a variety of reasons gelatin gels have been more studied than the fermented foods and this is mainly because of its use in photographic emulsions rather than its minor uses in food products. The much more widely relevant globular proteins had to wait until the attempts by many companies to make 'textured' products, ranging from lumps to filaments or 'fibres' for food use, in the 1960–70 period. In fact a substantial industry making textile fibres from proteins like casein and soy had already come into existence and been eliminated by competition from synthetics such as nylon, well before this. In view of all these applications it is remarkable how little was and still is known about fundamental aspects of protein gelation.

At a Faraday Society meeting in 1974 on protein gelation, it was clear that there was no consensus view, but that this was probably because different proteins formed gels in distinctly different ways. In a follow-up meeting on the same subject in 1982 there was more agreement and it is now possible to give a reasonable account of the processes involved.

Gelatin, for example forms gels that melt on heating, i.e. are thermolabile, and the gelatin molecule has a very high axial ratio. Globular proteins on the other hand form gels when their solutions are heated, and will only do this at relatively high concentrations. This alone is sufficient to suggest that different mechanisms must be involved.

Globular proteins when heated tend to form structures approximating to the random coil, and some early proposals assumed that extended chains would form and make the basic mesh structure of the gel, probably linking by hydrogen bonds. However this is very unlikely to be the structure of thermostable gels, though gelatin gels may be something like it (Fig. 3.5). Firstly the random coil is, as we have seen, a more or less spherical particle. Extended chains simply do not occur: even under conditions of high shear as in some filament extrusion processes there is no evidence for extended chain formation. In most cases intra-chain disulphide links will maintain a roughly spherical structure. (In the terminology of hydrodynamics, axial ratios of up to five are regarded as roughly spherical. Hydrodynamics provides a fuzzy picture of the molecule.)

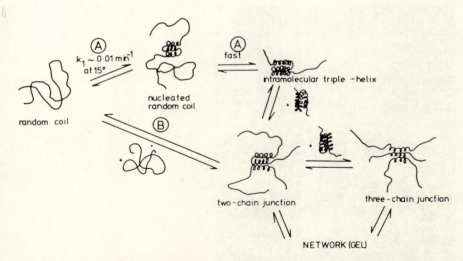

Fig. 3.5 Some possible conformational changes and interactions in forming the α-gelatin gel network. (From E. Finer, F. Franks, M.C. Phillips and A. Suggett (1975) *Biopolymers* **14**, pp. 1995–2005.)

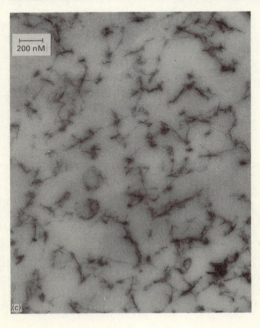

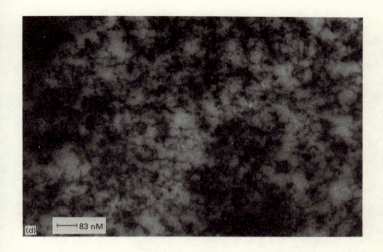

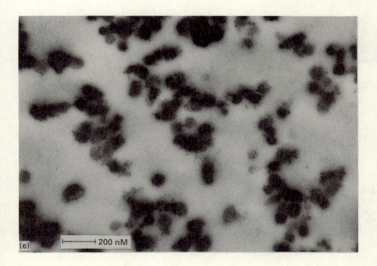

Fig. 3.6 Some protein gel structures typical of those found in many foods. Electron micrographs of stained sections. (A) A soy protein fraction at 15% total protein in water, gelled by heating. (B) The edge of a stretched filament, made from groundnut protein; about 20% protein. (C) An actomyosin gel, at about 5% protein, made by heating an actomyosin extract. (D) A soy protein gel, made from total crude soy protein at about 20% protein, individual molecules of storage protein, at about 8 nm size can just be made out. (E) A casein gel, made by lactic acid addition to a milk concentrate, at about 9% protein. Magnifications vary and are indicated by bars.

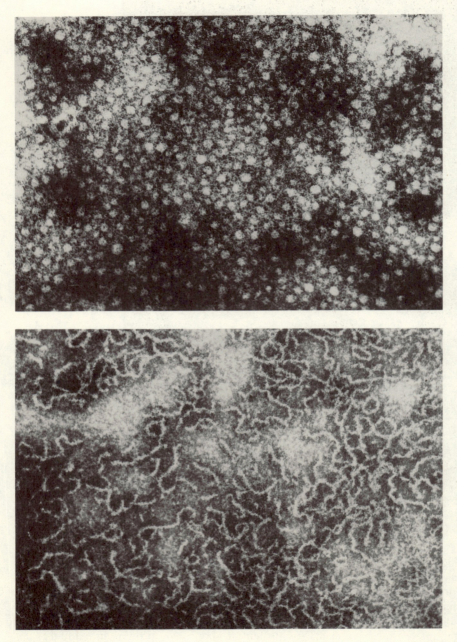

Fig. 3.7 Electron micrographs of bovine serum albumin aggregates. (Top) the result of heating at 100 °C for 8 s; (bottom) 69 °C for 30 min; 3% protein, negative stain on a carbon grid.

Mainly Proteins: Proteases, Gels and Fermented Foods

The first evidence on the structure of gels came from electron micrographs which showed that the fundamental mesh of the gel was made up from aggregates of spherical particles corresponding in size to the protein molecules. Some typical examples are shown in Fig. 3.6. This, the 'string of beads' model, has now been observed for many different globular proteins. It has also been found in extruded filaments, which in structure appear to be cylindrical gels. The strands of the gel—the term 'strand' is used rather than 'chain', to avoid inevitable confusion with the peptide chain—form during heating and tend to be irregular in shape. This is of course exactly what would be expected from our present knowledge of the behaviour of globular proteins in solution but leaves many questions unanswered. Why is it that some systems gel, while in slightly different conditions they coagulate? This, which can be a crucial matter for a food manufacturer depends in unknown ways on the aggregation mechanisms. Figure 3.7 illustrates the nature of the problem. When heated at 69 °C albumin forms mainly long high axial ratio aggregates, but as the temperature is raised to 100 °C, more and more spherical aggregates form. It is not surprising that albumin tends to gel at 69 °C but coagulates at 100 °C, but why it should aggregate in these two different modes is not known. Presumably the detailed surface structures and kinetics of interaction are different at the two temperatures and since the structure of albumin is now available it might give some clues. Similar effects might be expected for other proteins, including the seed globulins of primary importance, but have not been explored.

Apart from such alterations in the detailed aggregation mechanisms, in this instance due simply to temperature variation but also pH and ionic-strength dependent, there is a general concentration dependence under constant conditions.

With a protein like gelatin, and really it is unique, cooling a solution leads to either a gel or the appearance of a viscous solution if the concentration is not high enough (about 0.5%). It can be difficult to distinguish between a weak gel and a viscous solution. In this behaviour gelatin resembles many synthetic homopolymers and some carbohydrates such as agarose. Gelatin, which is a partial degradation product of collagen, forms a helix before the gel forms, and the interstrand links may represent partial reformation of the collagen structure. Figure 3.5 shows the current view of gelatin gel formation, based mainly on spectroscopic evidence. Globular proteins can form gels similar to this, in that they are thermolabile, but typically do so only in dissociating solvents such as 8 M urea. Under appropriate conditions the randomized chains can cross-link by mainly hydrogen bonds to form a mesh, and give transparent thermolabile gels at 5% protein. Soy protein will do this, and while there is no commercial interest in such gels, they are important as illustrating how the same protein mixture can be induced to make quite different kinds of gel by choosing the conditions.

Some years ago, when groundnut protein first became available as a result of the Groundnut Scheme, a test used for the quality of the protein was to make a slurry at about 20% protein and autoclave it in a can. Satisfactory protein produced an almost transparent firm gel; unsatisfactory protein gave opaque crumbling or even coagulated lumps. This remarkable transformation was a good predictor of usefulness in subsequent application, but it was never possible to find out what it was in the protein production process that led to this variability. It was clear however that the usefulness could be damaged by the kinds of process in use at that time and still in use.

QUANTITATIVE ASPECTS OF GELATION

We owe the first attempt at a quantitative treatment of aggregation to Von Smoluchowski, who described the process of random aggregation to form a spherical particle. Obviously aggregation in which any contact has equal probability of leading to adhesion leads to a spherical aggregate in the long run. The equation predicting the number distribution of different-sized aggregates is,

$$n_a = \frac{n_0(\beta n_0 t)^{a-1}}{(1+\beta n_0 t)^{a-1}}$$

where n_0 is the initial number of particles, n_a the number of number-average degree of aggregation a at time t. β is a constant.

Figure 3.8 compares the predictions of Smoluchowski with the actual size of albumin aggregates, and as might be expected it fails to predict the polymodal distribution actually found. This is because when albumin aggregates the basic

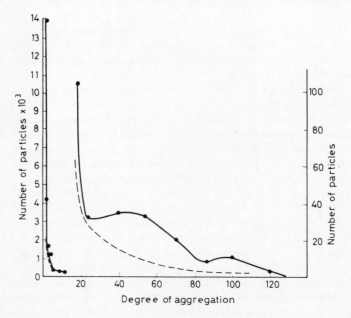

Fig. 3.8 Number-average degree of aggregation for albumin heated at 69 °C for 5 min, at 0.9% protein. The right-hand part of the curve is on the scale to the right. The approximate theoretical prediction from simple aggregation theory is shown (-----).

Mainly Proteins: Proteases, Gels and Fermented Foods

premise is not true. Nor is it likely to be accurate for proteins with a varied surface. So any deviation will tend to produce more asymmetric aggregates. The next development was the use of computer simulations to visualize the general shapes of aggregates when only a limited degree of aggregation was allowed. This produced a range of shapes, often highly asymmetric, which bore a strong resemblance to those seen in electron micrographs. We can therefore see that if there are sufficient of these to form a continuous gel mesh at the point when aggregation is stopped, a gel will result but if there is not then a coagulum is more probable.

Protein gels have pore sizes in the range from 100 to 300 Å, as judged from micrographs. Ogston found the relationship between concentration and pore size for a random array of fibres to be

$$p = \frac{1}{\sqrt{4\pi v L}}$$

where p is the average pore size, and v is the number of particles of length L. While he assumed that the fibres had length but no thickness, this does give a means of predicting the pore size of gels from the concentration, together with an estimate of the strand thickness from the string of beads-aggregate model. By a convergence, the same relationships are important in gel electrophoresis and filtration where a slightly different relationship was derived:

$$p = \frac{kd}{\sqrt{c}}$$

where k is a constant, d is the strand thickness and c the weight concentration. Here d, the strand thickness, is explicitly included. The culmination of this approach is shown in Fig. 3.9 where a predicted computer simulation of a protein gel is compared with a micrograph.

So it is possible, from the concentration and a model of the gelation process to predict the general molecular dimensions of the gel. There is as yet no simple connection between these and the attributes of protein gels important to food manufacturers but the way is now open to controlling the mechanisms of gelation to make a range of structures wider than hitherto available. Proteases will have an important part to play in this.

Some Fermented Foods

CHEESE MANUFACTURE

In reviewing the use of proteases in food manufacture cheese must take first place. While goat and ewe milk is used to make cheese the great bulk of it, about 2×10^{10} kg annually world-wide, is made from cow's milk. It is by far the largest user of commercial proteases, and as long ago as 1874 Hansen offered for sale a standardized rennin.

The first step in cheese manufacture is to add a culture of lactic acid bacteria and an extract of calf stomach containing the protease rennin. This enzyme has been

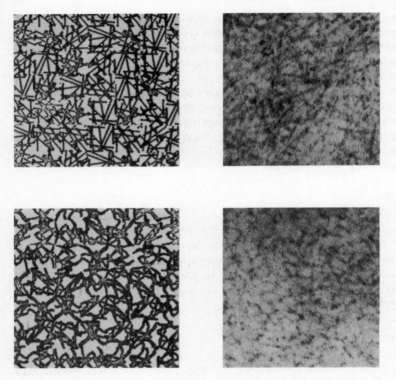

Fig. 3.9 Comparison of computer-drawn prediction with electron micrographs for (A) gels of lysozyme at pH 2.0 and (B) pH 3.0 α-chymotrypsin gel, both of which tend to form short rod-like aggregates under these conditions. (From A.H. Clark, F. Judge, J. Richards, J.M. Stubbs and A. Suggett (1981) *Int. J. Protein Peptide Res.* **17**, pp. 380–392.)

renamed chymosin (aspartic proteinase, EC 3.4.23.4), and this name will be used from now on. Chymosin occurs in the abomasum of the calf and clots the milk, no doubt as part of the digestive process. The bacteria bring about a reduction of pH by making lactic acid from lactose, and a combination of pH drop and chymosin action precipitates the casein. The precise nature of the coagulum varies between different cheeses and depends mainly on the pH at which separation occurs.

There are four different caseins—alpha-1, alpha-2, beta and kappa. All of them have genetic variants, with different incidence in different cattle varieties. This has little effect because of the practice of pooling milk from very large numbers of individual cows for manufacturing purposes, but could lead to variability if only single herds are used. The cDNA have been obtained and recently the genomic structural genes for rat casein have been characterized. The hope is to find the regions responsible for hormonal regulation.

Casein is a phosphorylated protein (on the serine residues) and kappa-casein is important, along with Ca^{2+} in stabilizing the protein in colloidal micelles in milk. Chymosin splits a single bond between methionine and phenylalanine at position 105 in kappa-casein. This is sufficient to destabilize the micelle leading to coagulation. The coagulum is allowed to contract, often with the aid of mechanical presses and the curd forms the characteristic structure of the cheese variety. Lipids, and some of the whey remain trapped, while the rest is drained off. Other proteases, and lipases originating in the milk itself as well as from the bacterial culture now slowly bring about changes in the amino acid, small peptide, and free fatty acid content, called 'ripening'. In cheddar fission of the Phe23–24 link in α-casein releases peptides thought to be crucial to the characteristic flavour. Proposals to modify this, and similar residues to assist in hydrolysis have been made but not yet put into effect. The longer it continues, and periods up to a year are typical, the more intense the flavour.

Chymosin supply depends on the slaughtering of calves, and while there should be a rough equivalence between calf numbers and milk production, in practice the supply of chymosin is insufficient. Chymosin shows a remarkable specificity and other proteases, such as pepsin, are less suitable because they split too many bonds. This not only results in loss of yield—peptides tend to be lost in the whey—but also the peptides are of the wrong sort. They have unwanted flavours. At least half of cheese made in the USA is now made with proteases secreted by *Mucor miehei*. It produces clotting similar to chymosin, but continues to act during the ripening period, and is more stable than chymosin, which loses its activity quickly. It is not a completely satisfactory substitute, and is used more for short ripening cheese, and vegetarian cheese which has recently become popular. A destabilized miehei protease has recently been made by genetic manipulation methods and is to be produced commercially.

However the chymosin gene has been expressed in the yeast *Kluyveromyces lactis*, and in *Aspergillus nidulans*, and secreted into the medium. It has been claimed that in practice *Aspergillus* is the best source, and that the yeast version can lack activity. In the near future this supply will fill part of the demand. Recombinant DNA-based chymosin has been approved as a food additive by the USA Food and Drug Administration. It will be interesting to see if it is acceptable for vegetarian use. This enzyme may not entirely replace the stomach extract, because this contains traces of other proteases and lipases, which have marginal effects during ripening. One of them, lingual lipase (EC 3.1.1.34) is known to originate in the calf's tongue, and is specific for shorter chain triglycerides found in milk. Rat lingual lipase has been cloned and is potentially available. Extra lipases of fungal origin are added to accelerate ripening in some cheeses, and it may be that some interesting new cheeses will be developed based on yeast-made chymosin together with added enzymes. The traditional cheeses are the outcome of hundreds of years of empirical experiment and will not be easy to improve on.

Whey

Whey disposal is properly seen as part of cheese-making. Most of it is thrown away, but increasingly, because of high BOD demands cannot be got rid of cheaply. There is pressure to develop a useful product based on it. It is not a promising starting material. It is essentially a dilute solution of lactose and the whey proteins, β-lactoglobulin and α-lactalbumin. The latter do have quite good properties for food use, and can replace egg-white in some applications. Unfortunately egg-white is cheap and plentiful, so that the relatively expensive processing of whey to give dry protein, and lactose, or a galactose–glucose mixture with the aid of galactosidase from yeast is difficult to sustain. As relative costs change, especially the cost of disposal, whey processing may become more prosperous. It is one of the few examples of the large-scale use of membrane filtration, to partially separate lactose from protein and water.

Other Enzymes

Some other minor uses of enzymes in milk are the addition of lysozyme as a bacteriostat in some baby-milk formulations, and the use of immobilized lactase from *Aspergillus*, as a means of removing lactose. Some populations, most notably those without indigenous dairy industries, cannot digest lactose, and lack the enzyme.

YOGHURT

Yoghurt depends, as does cheese, on a casein cross-linked gel for its structural properties, but differs in being more dilute, and hence still contains the whey or components deriving directly from it. The base composition of yoghurt is achieved either by removing water, by evaporation or by adding protein from another source. Table 3.2 shows some of the numerous different protein-rich powders which can be used to raise milk to the level required. Commercially, yoghurt is made to contain 14–15% milk solids of which, in the most popular low-fat yoghurts, about 10% is protein. Skim milk powder is usually used and it should be said that most of the potential additives given in Table 3.2, while they have all been shown to be capable of giving yoghurts acceptable to the consumer, have not found commercial use for a variety of reasons. Food-grade groundnut protein and leaf protein are not readily available, while soy protein can cause flavour difficulties.

Water is removed from milk by evaporation, which can also remove volatiles which are said to be undesirable in goat milk. Water removal for this purpose is also another possible field for the use of membrane filtration.

Yoghurt is made by the action of a mixture of *Lactobacillus bulgaricus* and *Streptomyces thermophilus* on the enriched milk. The key to commercial yoghurt manufacture lies in having just enough protein present to form the required gel, as a result of the casein precipitation brought about by their action. In marketing terms, yoghurt is actually a way of selling fruit, and profitability is dependent on a good cheap source.

Table 3.2 Composition of protein sources which might be used for yoghurt fortification*

	Composition (g/100 g)				
Source	Water	Protein	Fat	Lactose/carbohydrate	Ash
I. Animal (dairy)					
Milk powder					
Whole	3.0	26.5	27.0	37.5	6.0
Skimmed	3.0	35.5	1.0	52.0	8.5
Whey					
Acid	3.2	13.0	1.1	72.5	10.2
Sweet	3.0	13.0	1.0	74.0	9.0
High protein	4.0	75.0	11.0	8.0	2.0
Demineralized	3.0	13.0	0.8	82.4	0.8
Casein					
Commercial	7.5	88.5	0.2	0.0	3.8
Co-precipitate	4.0	83.0	1.5	1.0	10.5
Rennet	9.0	82.0	1.0	0.2	7.8
Acid	9.0	83.7	1.3	0.2	5.8
Sodium	5.0	89.0	1.2	0.3	4.5
Buttermilk					
Acid	4.8	37.6	5.7	44.5	7.4
Sweet	3.8	34.3	5.3	49.0	7.6
II. Soy bean flour					
Full fat	8.0	36.7	20.3	30.2	4.8
Medium fat	8.0	43.4	6.7	36.6	5.3
Defatted	8.0	47.0	1.0	38.0	6.0
III. Miscellaneous					
Peanut flour (defatted)	7.3	47.9	9.2	31.5	4.1
Leaf protein					
Whole concentrate		50–65	15–30	5–20	0.1–1.5
Solvent-extracted		65–80	2–5	10–30	2–8
Cytoplasmic fraction		70–95	0.5–5	3–30	0.5–1

*From R.R. Robinson and A.Y. Tamiye (1986) *Developments in Food Proteins*. Ed. B. Hudson, Elsevier Applied Science, Barking.

Milk used for yoghurt manufacture is usually heated and cooled again before inoculation, and this results in a coagulation mechanism different from that in cheese. It is known that the effect of heating is to cause the whey proteins, particularly β-lactoglobulin, to form filamentous aggregates attached to the casein micelles. These then become involved in the casein aggregation process and seem to assist in the formation of a smooth gel. If the milk is not pre-heated, a different and less desirable gel is obtained. It is not surprising that the use of different proteins as fortifiers brings about different end results in the nature of the gel.

Most of our information about protein gels comes from electron micrographs. Figures 3.6 and 3.10 give a good impression of what a casein gel is like. It is a typical bead structure, though the beads are relatively large since casein micelles are large, and they remain as the fundamental unit of the structure. In commercial processes involving the use of proteins as gel formers, it is always crucial to find the minimum protein level at which gels, as opposed to coagula, can form. This is done empirically, and the commercial yoghurt formulation represents the lowest level giving consistent results.

Lactobacillus bulgaricus produces the enzymes which attack the casein, as well as making lactic acid, and is the main gel forming agency. The *S. thermophilus* appears to be more concerned with peptidases and amino acid formation. They also make a variety of flavour components. Some strains are able to make polysaccharides from lactose, and this will become involved with gel formation, as will the bacteria themselves. Yoghurt is often pasteurized, but nevertheless contains large numbers of bacterial cells which must be at least supported by the gel structure.

There are obviously possibilities for biotechnology in yoghurt manufacture. A more controlled fission of casein to optimize gel formation, perhaps by modifying the proteases secreted by lactobacilli, and conceivably the modification of casein itself to make it a better structure former, are feasible. It is now possible to modify milk, by using transgenic animals with specific secretion into it. Success in all this will depend on a clear understanding of the way in which the milk proteins form structures in dairy products, of which cheese and yoghurt are the most important.

FERMENTED FOODS BASED MAINLY ON LEGUMES

In much the same way as every locality in Europe developed its own variety of cheese, all over the Far East, fermented foods of enormous variety both of composition and names are also made. Table 3.3 lists some of these where proteases play a significant role. In an arbitrary way, it combines a selection from the Himalayas with some perhaps rather better known ones from Japan. It illustrates an important point about the food industrial aspect of this kind of consumable. When interest in soy-based foods increased in Europe and the USA during the 1960-70 period, it was natural to visit the Far East to see how the local food industry made these products. It rapidly appeared that there was no food industry, larger than what would be called cottage-industry in Europe. Thus while the traditional methods have been much studied, large-scale processing is largely a Western development. At the same time as interest in fermented foods based on soya was raised, a great deal of development work on the use of soy protein to make textured foods by direct physical processing also took place. Very little in the way of marketable products emerged from a great deal of effort. In the course of it, however, it was confirmed that there are hazards attached to the use of soy protein. It contains a trypsin inhibitor, and a haemagglutinin, both of which must be modified in some way before consumption. The proteolysis inherent in the fermented production methods appears to be a very good way of doing this and is a

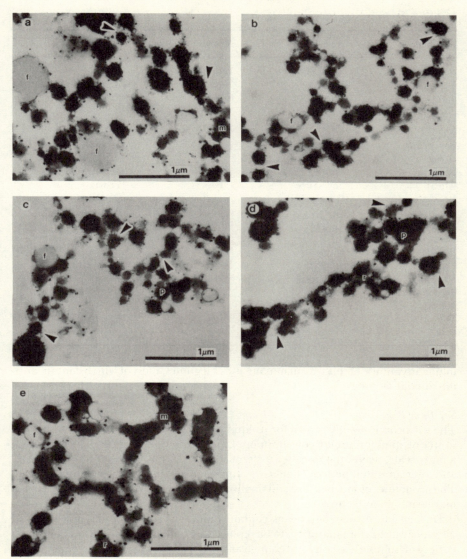

Fig. 3.10 Electron micrographs of casein micellar aggregates in yoghurt. (A) and (D) contain added skim milk powder. (B) and (E) were concentrated by reverse osmosis. (C) was concentrated by evaporation. All the structures, which show considerable variation in cavity size, and 'consumer-sensitive' rheological characteristics, are clearly built from the same fundamental casein micelle units. (From R. Robinson and A. Tamime (1986) *Developments in Food Proteins*. Ed. B. Hudson, Elsevier Applied Science, Barking.)

Table 3.3 Some fermented foods

Raw material	Names	Region	Organism
Soy beans	Tempeh	Indonesia	
	Kinema	Nepal	*Aspergillus* sp.
	Bhari	Sikkim	*Rhizopus* sp.
	Sufu	S.E. Asia	
	Natta	Japan	
Soy beans + rice	Miso	Japan, USA	*Aspergillus oryzae*, *Aspergillus sojae*
Soybeans + wheat	Shoyu	Japan	*Lactobacillus* sp.
	Soy sauce	UK	Yeasts
Leaves of *Brassica* sp.	Gundruk	Nepal	(?) Yeasts Bacteria
Radish roots, cauliflower	Sintri	Himalayas	
Bamboo shoots	Mesu	Nepal	
Coconut	Semeyi	Indonesia	*Bacillus* sp.
Rice flour, bananas and honey	Shel roti	Sikkim	Yeasts
Yak or cow milk	Churpi	Tibet	*Lactobacillus*

major advantage of this way of making use of legumes, most of which contain anti-nutritional factors.

Soy Bean Processing
The soy bean is mainly grown for its lipid content, which is still the largest single source of lipids for human consumption. Lipid extraction, these days by hexane on a large scale, leaves the so-called 'soy meal', which contains about 50% protein, and after an initial period when it was thrown away now sells as an animal feed. The economics of soy meal are still dominated by the fact that it is a by-product of margarine manufacture.

During maturation, soy beans contain numerous starch grains, but these all disappear just before maturity, to be replaced by lipid. In addition, soy contains a variety of 'soluble' carbohydrates, such as stachyose, arabinosans and galactosans. These, together with some sucrose are the substrate on which organisms grow during fermentation. The protein is predominantly contained in protein bodies and is a mixture of seed globulins closely similar to those found in all legumes (see below). Figure 3.11 shows a section of a mature soy bean. The protein bodies survive almost untouched, including their membrane, into the defatted meal. There is almost certainly scope for the use of enzymes in protein extraction methods, and it would be interesting to try the effects of both cellulases and phospholipases. Two further derivatives are commercially available. Enriched, washed or concentrated meal is made by washing out the soluble carbohydrates at pH 5, which leaves most of the protein behind. Secondly, protein isolate is available, and is made by extracting the meal with water at pH 8, which dissolves most of the protein, followed by precipitation at pH 4.6. This material is better than 90% protein and is the only legume protein currently available in tonne quantities. It is sometimes described as 'soluble' in commercial practice. Most of it is not, by strict definition, and is capable of forming more or less stable colloidal dispersions only. It resembles casein, and is indeed known as 'soybean casein' in European customs documents. A useful practical test is the ability to withstand sedimentation at $100\,000 \times g$ for 30 min which truly soluble protein will do. Most soy protein is aggregated during defatting, not as might be thought by the action of hot hexane, but by the common practice of removing the last traces of hexane with steam. It is possible to make meal containing soluble protein, and extract it, but this requires special procedures and only a tiny proportion of soy is processed in this way. Food use also requires a series of other precautions. The outcome is that while soy protein, soluble and fit for human consumption, is available, it is rare, and approximately five times as expensive as the cruder materials used for animal feeds. It is not, and never has been, a cheap protein in isolated form, though it is much cheaper than meat proteins, which are even more expensive, but have very different properties.

Soy beans for fermentation start by a soaking and grinding, followed by decantation of the excess water. This removes a good part of the soluble carbohydrate. The soluble carbohydrates are known to be the 'flatulence factor' which makes eating untreated soy meal a somewhat explosive experience. No product containing them could safely be put on the market. The residues are eliminated by the growth of the organisms. In non-fermentative processing they are now so difficult to dispose of because of high BOD, that they make soy-meal processing uneconomical in Europe, or anywhere else where strict controls are maintained. So far no-one has found a use for them.

The next step is to sieve and remove more water by pressing and then heating to 100 °C. Cottage processes might avoid this step, which is intended to remove unwanted organisms. It also makes the slurry more amenable to enzyme attack, for the reasons already discussed. Calcium and magnesium salts are sometimes added at this stage as coagulants. All the seed globulins are very rich in glutamate residues, which react with added Ca^{2+} to release H^+, so that the pH drops sharply. Similar coagulation can be obtained simply by acidifying, but the necessary acids like hydrochloric or sulphuric, which would be used in large-scale processing, are less easily come by in the Himalayas.

The product is a coagulum of protein, lipid and cellulose cell walls, which is then inoculated with the appropriate organism for the fermentation. In Nepal, a dry cake called *murcha* seems to be used as a general-purpose starter. It contains numerous strains of yeast, *Rhizopus* and *Mucor*, which remain viable for years. Its production is in the hands of one caste, who make it from rice flour, together with the roots of plumbago and leaves of *Buddleja* species.

In some cases, the sequence includes an alkaline extraction of the meal, followed by precipitation of the protein, which would give a coagulum containing less cell-wall material. This would yield a smoother and stronger gel. *Tempeh*, which is

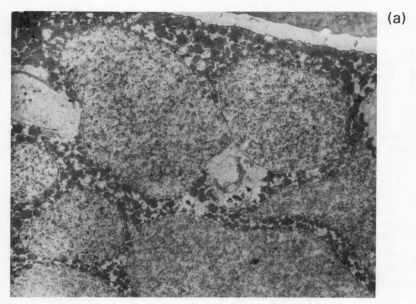

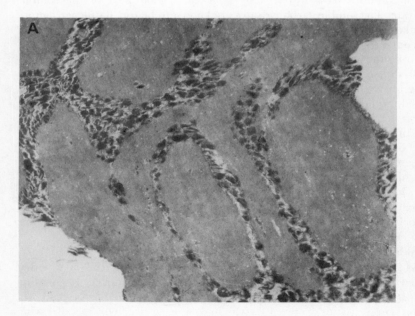

Fig. 3.11 An electron micrograph of sections of soy beans, originating in Nigeria (N) or the USA (A). The difference in appearance may reflect differences in moisture content. (B) Photomicrographs of defatted meal, in dilute osmic acid, to show the intact protein bodies. SEM images of soy beans, at the top, after defatting and once again reveals the intact protein bodies. At

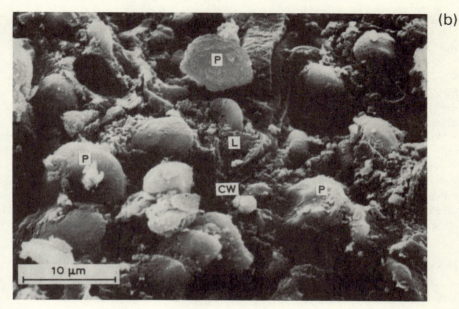

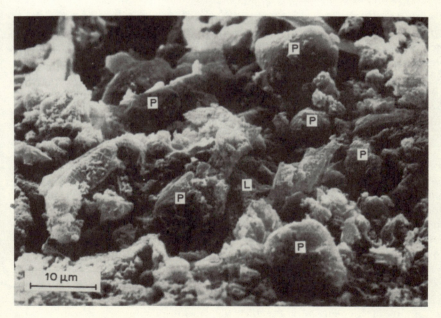

the bottom is the result of germination before defatting where the process of protein body breakdown has begun. Key: P = protein body; L = lipid matrix residue; CW = cell wall. (From E. Wassef, G.H. Palmer and M.G. Poxton (1988) *J. Sci. Food Agric.* **44**, pp. 201–214.)

simply water-extracted bean, relies to a considerable extent on fungal mycelia to give it a desirable texture. All these products are known to depend on the details of the washing and heating process for the quality of the gel formed.

Soy beans contain phytic acid. This, which is a mixture of phosphorylated inositols, forms a complex both with the protein and added Ca^{2+}. It is hydrolysed by phytases during fermentation, which is an advantage. In a few cases, Ca^{2+} deficiency has been ascribed to diets containing excess phytic acid (in unleavened bread as well as in seeds) since it makes Ca^{2+} difficult to absorb from the gut.

As large-scale methods for making soy-fermented products develop, and so far only *miso* and *shoya* have been attempted, it appears that there will be developments of the existing soy processes. In fact the enriched meals, and protein isolates developed during the 1960s, which have equivalents in the cottage industries of the East, are clearly suitable substrates for fermentations. (There is said to be a pilot plant in Singapore which uses immobilized enzymes in columns for speeded-up manufacture of soy sauce.) There is every indication that they will take their place in the supermarket.

Proteases

Proteases are specialized esterases that attack peptide links in their normal activity though they will usually attack simple esters as well. Proteases are extremely widespread and rather remarkably the activity is shown by at least four families of enzymes that appear to have evolved independently. These are the serine proteases, the cysteine proteases, the aspartate proteases, and the metallo-proteinases, which require a Zn^{2+} or Mg^{2+} in the active centre.

A further distinction is made into endo-proteinases, which split peptide links in the body of the chain, sometimes with marked specificity requirements for amino acid residues, and exo-peptidases which remove either the C-terminal or the N-terminal residues, again with some specificity for certain residues. The exo-peptidases are metallo-enzymes.

In their normal function, which extends from digestion to the most detailed processing of newly synthesized chains in the cell, the proteases display a great range of specificities. Some of them, typically the thiol enzymes of plant origin like papain, have a wide specificity. Others seem to be designed to attack just one peptide link to bring about a defined change in behaviour. The membrane proteases, that remove leader sequences as peptide chains pass through, are good examples, as are the proteases involved in blood-clotting mechanisms, or the example already quoted of chymosin on casein in milk clotting.

DIGESTIVE ENZYMES AND PROTEASE INHIBITORS

The human digestive enzymes, of which the most important are trypsin, pepsin, chymotrypsin and a group of carboxypeptidases, are as a result of their concerted action able to break down most of the protein in the diet to amino acids and small peptides. This is what is found in the mesenteric portal vein. It is important to realize however that single proteases acting alone have only a very limited attack on dietary protein. Figure 3.12 shows the course of the activity on some defined

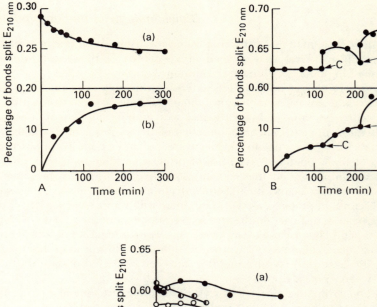

Fig. 3.12 The extent of bond fission obtained with some typical mammalian digestive proteases on a variety of substrates, of which edestin is a typical seed globulin, and albumin a fairly typical 'meat' protein. Fission of peptide bonds brings about changes in the absorbance (E) at 210 nm, which was also monitored. (**A**) Tryptic hydrolysis of protamine at 37 °C; 0.2 mg trypsin/ml and 5 mg protamine/ml. (a) Change in E at 210 nm; (b) extent of peptide bond fission as shown by formol titration. (**B**) Enzymic hydrolysis of bovine albumin at 37 °C. The initial enzyme–substrate mixture contained 10 mg trypsin and 500 mg of albumin, in 50 ml buffer. At point C 10 mg chymotrypsin was added; at point P the pH was adjusted to 1.9 with 3 drops conc. HCl and 10 mg pepsin was added. Increases in E at points C and P are due to the added enzyme. Turbidity appeared after the addition of pepsin. The lower curve shows the extent of bond fission, determined by formol titrations. (**C**) Peptic hydrolysis of ○, edestin, ◐, ovalbumin and ●, bovine albumin at 37 °C. (a) Variation in absorption at 210 nm. Substrate conc., 10 mg/ml. For bovine albumin, 1 mg pepsin/ml; for the remainder, 0.5 mg pepsin/ml; (b) extent of peptide-bond fission determined by formol titration. (From M.P. Tombs and N.F. Maclagan (1962) *Biochem. J.* **84**, pp. 1–6.)

proteins, and only a small proportion of the bonds are split. It is extremely difficult to obtain complete hydrolysis of proteins with proteases. (If it were easy, it would be very useful in amino acid analysis technique.)

It is customary to use small synthetic esters in protease assays, and while these are quite acceptable ways of finding the amount of protease present, two different proteases with the same activity against a synthetic substrate might have completely different activity against a mixture of dietary proteins. The activity in standard assays is a poor guide to the effectiveness of the enzyme in application. It is because of the cooperative activities of proteases in digestion that protease inhibitors are important in the food industry. Legumes, even such staples as potatoes, contain specific trypsin and chymotrypsin inhibitors that are themselves proteins. Clearly, in practice, a trypsin inhibitor will prevent the activity of exo-peptidases because these act on the terminal groups exposed as a result of tryptic attack. Exo-peptidases are not exempt, and *Actinomycetes* make a range of peptides that inhibit them. The effects of protease inhibitors do not stop at impaired digestion, however, and trypsin inhibitor has been suspected of damage to the pancreas, where trypsin is made.

It would be wrong to suppose that trypsin inhibitors are the only, or even the major, toxic factors in seeds. For example a study on more than 100 varieties of soy beans, with a wide range of trypsin inhibitor activity, were also assayed for their PER (protein efficiency ratio, a measure of the ability of rats to use the protein) and there was no correlation at all between these two variables. There was however a negative correlation between PER and pancreatic hypertrophy, suggesting a less direct effect than one simply on digestive enzymes.

Protease inhibitors have been found in a huge range of raw materials including all the legumes, common cereals and eggs but fortunately lose their activity as a result of cooking, or paradoxically as a result of protease attack in fermented foods. An important, but beneficial inhibitor is the α_1-antitrypsin of the blood, which is also an elastase inhibitor. It is thought to play a role in protecting the lungs against bacterial protease attack. About 1 in 400 of the European population lack this protein, and usually show lung-damage symptoms.

Contrasts can also be drawn between the efficiency of the concerted action of a mixture of enzymes in the mammalian gut, with each enzyme of limited specificity, and the action of a single enzyme such as subtilisin, secreted for digestive purposes by *Bacillus subtilis*, with a wide specificity. The mammalian system is effective because it is a combination of endo- and exo-peptidases. It follows that in applications where extensive digestion is sought as a result of adding proteases, in food processing it should be advantageous that both classes of enzyme are present. A combination of a wide specificity endo-peptidase with an exo-peptidase should be particularly effective. Although endo-peptidases are readily available, no source of exo-peptidases for industrial use exists. This may be because no bacterial or fungal exo-peptidases have been characterized, since nearly all the industrial proteases originate in them. It ought to be possible to insert a procarboxypeptidase gene into yeast, to obtain supplies of such enzymes. The gene has been cloned.

It is common in some countries to take enzyme preparations as an aid to digestion, and there is a small specialist market for 'predigested foods'. The main

Mainly Proteins: Proteases, Gels and Fermented Foods

application however would be in modified 'fermentation' processes and in the production of special peptides with useful properties, perhaps as emulsifiers.

USES OF PROTEASES IN FOOD PROCESSING

Table 3.4 lists some established applications of proteases, and it is at once evident that the three main classes of proteases have all found a place. The exception appears to be the metallo-enzymes and the exo-peptidases. As already pointed out there are no readily available sources of such enzymes and they thus are one class where the resources of biotechnology might have most impact. Table 3.4 also lists some potential applications of proteases and the exo-peptidases would have a part to play in the situations where extensive digestion is needed, and some special requirements. Various collections of data are available giving the tonnages of enzymes used in different applications. The difficulty is that purified enzymes are rarely used and in most cases the enzyme itself is only a small proportion of the protein present, frequently $<1\%$. Comparative weights are of little value therefore. It is usual for enzymes to be sold on the basis of activity in some agreed assay system, and manufacturers will sometimes give an assurance that certain types of activity are absent. For example, an amylase might be sold with a guarantee that the protease levels are less than a fixed amount. On the other hand, identical material can be sold as a 'lipase' or a 'protease' depending on the customer's requirements.

Cysteine Ezymes: Papain

Papain appears several times in Table 3.4 and is one of the more nearly homogeneous enzymes available. Its major use, quantitatively, is in haze reduction in beer manufacture. Beer, when chilled forms hazes containing proteins which can be digested by papain, without otherwise affecting the quality of the beer. It cannot be used for the same purpose on wines, probably because the level of polyphenolics in wine is higher than in beer. Wine is clarified by bentonite filtration instead, but a modified protease might be found that would be more effective.

Another use for papain is in processes involving yeast autolysis. Addition marginally improves the yield. The extent to which this is actually done in commercial practice is obscure. The most widely known use of papain is in meat tenderizing. It is known that in the process that converts muscle to meat (and this conversion, although perhaps not what is normally thought of as a food-manufacturing process is best understood in this context), the essential changes are brought about by a mixture of proteases. The precise history of the muscle, as well as the type of muscle itself can have a large effect on the quality of the resulting meat. The basic idea is therefore to add proteases to try and improve on those already present. Although this has been described as tenderizing the objectives can go somewhat further. Texture is a more complex matter than simply tenderness, while flavour can be influenced by the presence of amino acids and peptides.

Attempts so far have been based on the use of papain, no doubt because it was the only protease available for food use in reasonable quantity. They have not

Table 3.4 Uses of proteases in food processing

Function	Enzyme	Type
Brewing: haze reduction in beer	Papain, bromelin, ficin	Cysteine
Baking: dough modification	Papain, *S. griseus*, *Aspergillus* proteases	Serine
Cheese-making	Chymosin, pepsin, *M. miehei* protease Penicillopepsin ex. *Penicillium* sp.	Aspartate
Protein hydrolysates for flavours and in fermentations	Subtilisin, from *B. subtilis* and from *B. amyloliquifaciens*, *Aspergillus* Alkaline protease Collagenase	Serine
Muscle to meat	Cathepsins, papain	
Emulsifier production	Proteases from *S. fradiae*, *B. lichiniformis*, 'pronase' and 'pancreatin'	?
Yeast autolysis	Papain	Cysteine
Aspartame synthesis	Thermolysin	
Inhibitor removal	Modified wide spectrum	

been particularly successful, since the quality of the treated meat was not seen as good enough. The initial attempts at selling papain-treated meat invited comparison with good quality steak, and the public was not impressed. Current usage is more confined to manufactured meat products.

A process which involved injecting papain in solution into the vascular system of animals just before slaughter, followed by activation of the papain by the post-mortem drop in the muscle pH, is unlikely to be used. It is not without problems but would almost certainly be held to be unacceptable in most countries. The process has been a relative failure because papain splits too many peptide bonds (and probably the wrong ones). The outcome is soft and lacking in the structure seen in the target model of fillet steak. It is now known that specific proteases, particularly a Ca^{2+}-activated protease, and collagenases, are involved. These enzymes split, with high specificity, a limited number of bonds in some minor components that produce the desired structure. Once more biotechnology offers the opportunity to obtain supplies of these enzymes, with a good chance that they will give better results than papain. So the conclusion must be that while the field of meat tenderizing by proteases is commercially quiescent at the moment, it could reappear with much better prospects of success.

Papain is potentially of use in almost every situation where proteases might be used and is known to affect the properties of dough in bread-making, the malting

Mainly Proteins: Proteases, Gels and Fermented Foods

process in brewing, and can be added to almost any fermentation to assist the *in situ* enzymes. It has one advantage as a thiol enzyme, in that oxidizers can be used to eliminate its activity fairly easily. It is often important in protease applications, and the best-known example is cheese-making, to be able to stop the fission from going too far.

Although pineapple and figs both provide a source of very similar enzymes, commercially the source of papain is the latex of the papaya (*Carica papaya*) grown for the purpose in the tropics. The latex is dried, and consists of a mixture of closely related forms of papain, chymopapain and some lysozyme. For food use it is refined to remove insects, fungal contamination and so forth. There appears to be no point in developing alternative sources.

Aspartate Enzymes
Chymosin appears to be the only example of this group, together with minor use of the related pepsin. Its role in cheese manufacture, and its production by biotechnology means, has already been discussed.

Serine Enzymes
A large proportion of the fungal and bacterial enzymes that have found use, albeit in crude fermentations, belong to this class and subtilisin can serve as an example. Because it is important in non-food use—it is used in enzyme detergents in a far greater tonnage than any enzyme used for food use—it has been the subject of much research. The detailed structure of subtilisin has been known for some years.

Figure 3.13 shows a stereographic view of the active site, with a synthetic substrate *in situ*. The catalytic activity of the enzyme is dependent on the three residues—serine-221 which is acylated, His-64 and Asp-32.

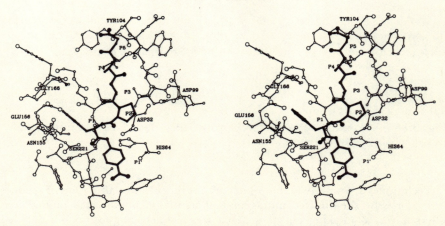

Fig. 3.13 Stereographic view of subtilisin containing N-succinyl-Ala–Ala–Pro–Phe–*p*-nitroanilide in the active centre. Important residues, particularly the catalytic triad are labelled. To see the figure in depth, cross the eyes until the images coincide. (From P. Carter and J. Wells (1988) *Nature* **332**, pp. 564–66.)

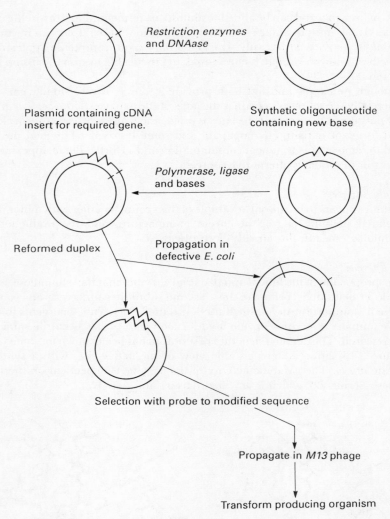

Fig. 3.14 An outline of the method for inserting modified DNA into a gene to bring about directed mutagenesis.

Techniques are now available which permit the selective alteration of residues in enzymes by site-directed mutagenesis. Potentially, site-directed mutagenesis offers a way of making enzymes much more closely aligned to the detailed requirements of applications than any that are available at present. The proteases were among the first enzymes to be modified (the first was tyrosyl tRNA transferase) possibly because they were among the first to have their structures and detailed mechanisms established.

Figure 3.14 gives an outline of the method by which selective alterations are made. The cDNA, contained in a plasmid, is partially digested with restriction enzymes designed to remove the stretch of DNA it is intended to modify. This is then treated with a synthetic DNA containing at least 16 bases and with a mismatch, designed to alter the required amino acid residue. This reacts with the complementary strand, sufficiently well to permit DNA polymerase and ligase to reform the strand. The crucial practical discovery that made this technique possible was the discovery of an *E. coli* strain that lacked the DNA repair mechanism. This (which is present in all cells) is able to make good and eliminate mismatches which arise from the presence of mutagens in the environment. The effect is to make it impossible to propagate the modified DNA except in the defective strain. Also shown in Fig. 3.14 is a more recent method which uses the *M13* phage to multiply the cDNA in much the same way as it is used to make DNA for sequencing purposes. Once the single-strand cDNA is made it is straightforward to make the complementary strand, this time with the modified triplet but no mismatch, and obtain production of the modified protein in the usual way.

In subtilisin the three important residues were replaced by alanine in all possible combinations in a series of experiments which were really concerned with how an enzyme requiring three residues for activity could have evolved by sequential single changes in the sequence. In fact, installing His-64 gave a marked increase, while addition of Asp-32 gave a further increase in catalytic rate, compared with the enzyme having only Ser in the triad. Substrate binding to the enzyme was also modified by minor changes in residues. Interestingly, converting the Ser to a Cys, making it in effect a cysteine protease, did not lead to much activity. The reason is thought to be that in the cysteine proteases, an Asn residue is used rather than an aspartate in the catalytic system.

Another residue that has been found to be important in subtilisin is Asn-155. Replacement by Thr, Gln, Asp and His all led to reduced activity. Like many esterases subtilisin is only active when the His is uncharged, and by changing charged residues in the vicinity to alter the pK of the histidine it is possible to alter the pH dependence a little. Subtilisin has a substrate-binding pocket containing a Gly-166. Substitution by any of 20 different residues changes the specificity against small substrates quite markedly. Similar experiments have been done on trypsin and by altering residues in the substrate-binding pocket, the specificity could be changed though not by a great deal.

The specificity of proteases appears to be determined by at the most two residues, and often only one. Trypsin for example will act next to any of the Lys or Arg residues. Proteases would be much more useful if they had specificities defined by four or more residues—rather like restriction enzymes acting on DNA—so that they could be used to attack far fewer but more carefully chosen bonds. It may be possible to do this by taking account of the shape of the substrate protein. One might even envisage a combination of antibody shape recognition with linked protease active sites.

A protease that would selectively attack the trypsin inhibitor of soy, for example, would be very useful. In a sense trypsin does this, since it forms a stable complex with trypsin inhibitor. Unfortunately, it would be necessary to add an equal weight of trypsin (they have roughly equal molecular weights) and since trypsin inhibitor is at least 10% of a typical soy protein isolate, even if trypsin should become available from yeast, it is unlikely to be used. What is needed is a protease designed to interact with a suitable part of the sequence. Even if it did not completely eliminate the inhibitor activity, it would make it more susceptible to heat inactivation. This will be even more useful in dealing with a number of legumes where the trypsin inhibitors, unlike soy, are heat resistant, and not eliminated by normal processing.

PARTIAL HYDROLYSIS

Several classes of proteases have been used in experimental investigations of the effects of partial hydrolysis on caseins and whey proteins, as well as soy and indeed all the commercially available proteins, and some that are not. Acid or alkaline hydrolysis has also been used, but there is an important difference when proteases are employed. The Gln and Asn residues are quite labile and release ammonia under surprisingly mild conditions. This does not happen with proteases.

Some peptides are good emulsifiers. Not surprisingly, these turn out to be the ones with mainly hydrophobic groups at one end of the chain, and hydrophilic ones at the other. Peptides with this character are likely to occur on a random basis in any mixture produced by partial fission, and the emulsifying power of almost any mixture of proteins can be improved in this way. Research is now going forward to make emulsifiers by fractionation of partial hydrolysates of cheap proteins. So far there are no significant commercial applications, and it is possible that these efforts will be overtaken by a different approach. This is to modify the synthetic apparatus of a suitable host to manufacture a carefully designed peptide directly. It might be possible to modify one of the caseins so that, after synthesis, it could be easily broken down to give the desired peptide, perhaps in the course of cheese-making. It would make whey more valuable than it is now. Alternatively, a seed globulin might be the best route. The problem is no different from that of inducing the synthesis of any protein in a host organism, but this may be an example where a wholly synthetic gene will have to be made.

Peptides can be undesirable both because of flavour and because some of them have pharmacological action not wanted in foodstuffs. For example β-casein yields the peptide Tyr–Pro–Phe–Pro–Gly–Pro–Ile–, which is an opioid. Crude hydrolysates run the risk of containing this kind of peptide, which should be absent from those made by directed synthesis.

Proteolytic fission also promises to be useful in a quite different context, as a way of altering the gelation characteristics of proteins. It can be used to recover damaged protein. Protein which has been aggregated and made insoluble, often by heat treatment, has impaired gelation capacity. Although it cannot be made soluble again by protease digestion, it can sometimes be induced to form a colloidal sol, simply because the average particle size is lower. This then will give more useful properties. Protease action on soluble protein will sometimes aid gelation, though the reasons are not clear. Soy protein treated with bromelin, gels, and this is thought to be because thiol groups were activated by conformational change.

Mainly Proteins: Proteases, Gels and Fermented Foods

While this is certainly possible so are many other mechanisms, and so far this approach is on an empirical basis.

Directed synthesis might also be used to produce proteins which could be gelled using the same kind of mechanisms as those that induce clotting as a result of specific bond fission in fibrinogen. However, such applications are unlikely to find commercial application in the foreseeable future, though their feasibility may be tested soon.

THERMOPHILES

There has been much research into thermophilic proteases, and proteases have been isolated from thermophilic organisms, and expressed in mesophiles. Thermolysin is the best known, from *Bacillus stearothermophilus*. It is a Zn^{2+} metallo-enzyme. Others are serine enzymes, and so far no thiol protease has been found. The main interest in these enzymes is for their stability at high pH and temperature in enzyme detergents. It is difficult to see what use they might have in food processing, since as is clear from several examples already discussed, instability is often a desirable feature of proteases used to make foods. They might however be of use in chemical processes for flavour components (see Chapter 5).

Other Covalent Modifications

Although proteolysis is one of the most important covalent changes that happen in foods, other covalent modifications to proteins can also occur, and many of these involve enzymes. We have already considered those that lead to cross-links.

PHOSPHORYLATION

Phosphorylation of certain tyrosines and serines by kinases is an important means of metabolic control. Whether similar enzymes are responsible for the attachment of phosphate groups to serine in casein in the mammary gland is not clear. However, such enzymes clearly exist and are capable of phosphorylating a wide range of proteins. They are not yet available, but a more serious drawback is that the substrate is AMP or ATP. It might be possible to find an enzyme that uses pyrophosphate, or even one that could exchange groups between casein and other proteins. Studies with chemically phosphorylated proteins show that they can be induced to gel by addition of Ca^{2+}. This was the case for β-lactoglobulin and soy protein. Microorganisms can use them, but no mammalian data are available. Phosphatase can remove the phosphate groups, and in non-aqueous media might even be used to insert them.

ACYLATION

There are a great many studies on acylated proteins. Groups ranging from acetyl to palmitoyl have been inserted, and as might be expected, the surface properties of the proteins is altered. Reaction is mainly with Lys groups, and thus the charge balance of the protein is also altered. They are said to have good foaming capacity but poor stability. They require extensive safety testing, and there appear to be no enzymes able to carry out the same reactions.

GLYCOSYLATION

The use of glycosylated proteins is a much more promising area. Glycoproteins are of course well known, and range from examples with only one or two sugar residues to those where more than half the weight is sugar side chains. The enzymes and synthetic pathways are relatively little known but offer great potential for making modified seed globulins. These, which are dominated by the storage globulins, contain little glycoprotein. Arachin and glycinin contain none. (There are glycoproteins present in seeds which have attracted much attention as lectins, but they are quantitatively minor components; see Chapter 6.) In fact the cheap bulk proteins which are currently available contain little glycoprotein. The egg-white proteins appear to be the only exception.

Chemical glycosylation of whey protein has shown that they have a range of useful emulsifying properties and while chemically modified proteins are unlikely to be acceptable for food use in the current climate of opinion, enzyme-based processes would almost certainly be feasible and lead to useful ingredients. Rather little is known about the enzyme systems that attach hexoses to proteins, and for the present it is not at all clear that they would be of general use. A triplet of Thr, Pro and Asn residues have been recognized as glycosylation sites in some lipases, and it is possible that site-directed mutagenesis might permit the insertion of glycosylation sites into suitable proteins. Commercial exploitation is many years away.

Conclusions

There is obviously great scope for the application of biotechnology methods to proteins and proteases. If only because it should be possible to modify both enzyme and substrate, in a coordinated way the exploitation of biotechnology ought to have more to offer in this part of the food industry than any other. At the same time, the applications seem to be some way off. The reasons for this may lie in the inability of the food manufacturer to specify what is required. If as seems probable, fermented legume-based foods grow in popularity, no-one will be able to make the connection between consumer preference and the use of a particular battery of enzymes in their manufacture except by laborious and expensive consumer trials. Consumer preference trials are not in fact very good predictors of commercial success. If they were there would be fewer failures of new product launches. Thus the prospect is of a period of rather empirical exploitation of what could become a numerous collection of modified proteases. Where the problem can be specified, rapid progress is to be expected. The production of chymosin in cheese manufacture is the outstanding present example, and it is obvious why this was the first to appear. Trials of specific proteases in meat processing, and proteases designed to attack known sequences in protease inhibitors are other examples. Inactivation of undesirable haemagglutinins may be achieved by proteases or by specific glycoprotein-degrading enzymes. Production of peptide emulsifiers is also feasible, though not yet shown to be commercially possible. Modification to improve nutritional status, by inclusion of better levels of essential amino acids, is not needed in developed countries, but could be used to improve their supply elsewhere. (This is a more complex problem involving plant nutrition and agriculture rather than food manufacture.) The supply of modified caseins offers the possibility of interesting new dairy products, and no doubt their sequences are being studied with this in mind. The sequences of seed globulins are now well known, and offer similar possibilities. There are limits, since induced alterations must lie within the amino acid production capabilities of the plant. Insertions of the gene for metallothioneine, with a high content of cysteine, are likely to outrun the sulphur metabolism of the seed. On the other hand, this protein would be interesting to the food manufacturer if it were available, in its effect on gel and structure formation.

It will be some years before foodstuffs based on modified proteins are on the shelves. Products made with novel enzymes are already appearing, and new ones are likely to continue for a long time.

Chapter 4

Cereals: Baking and Brewing

Introduction

In every part of the world cereals are the most important foodstuffs, with tonnages that far exceed any other source. An indication is given in Fig. 4.1 and confirms the predominant position of wheat, maize, rice and barley.

The cereals are the most important source of protein but also contribute enormous quantities of starch. We have already considered the uses of corn starch in Chapter 2, as a major source of high-fructose syrups. In this chapter we will be describing another, and much older use for starch. This too involves hydrolysis to mono- and disaccharides by amylases and amyloglucosidase, but the resulting sugars are converted to ethanol, rather than fructose. The process is of course brewing, and although it traditionally uses barley, any cereal can be used. Sorghum for example is widely used in Africa.

A large part of the cereal production is used for feeding animals, essentially for meat production. Most of the remainder is used for baking bread.

The nutritional status of cereals is, curiously for such a basic food, slightly deficient. As is indicated in Table 4.1, they are all lacking in one or more of the essential amino acids. Table 4.1 includes the composition of a legume, and it is clear that a combination of legumes and cereals is just adequate in essential amino acids. Just such a combination is the basic diet of very large numbers of people in developing countries, supplemented by much more limited sources of animal protein.

Most food processors, though not all, are located in developed countries where the normal diet is adequate. There is no significant malnutrition in Europe, Japan or North America due to deficiencies in the available food. Thus, food manufacturers are not usually primarily concerned to improve the nutritional

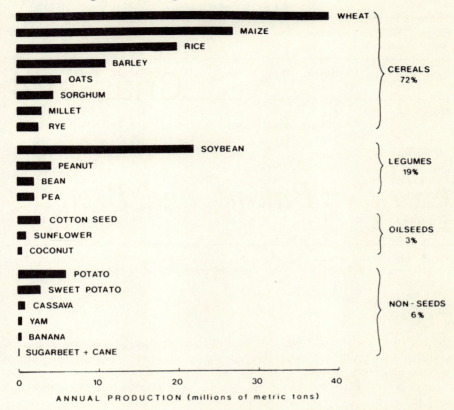

Fig. 4.1 Annual production of major protein and carbohydrate food sources, for 1980. (From P.I. Payne (1983) 'Breeding for proteins in food crops', *Seed Proteins*. Ed. J. Daussant, J. Mosse and J. Vaughan. Academic Press, London.)

value of what they sell. Nor are consumers, who know that if they continue to eat what is actually a highly varied diet, they will obtain all the necessary nutrients. In fact any one food product is usually such a small part of the diet that its effect on overall nutrition is negligible. There are always exceptions of course, and individuals have been found living exclusively on raw sausages, or carrots, with the inevitable consequences to their nutritional status. But most consumers (and extensive trials have been carried out), are concerned with flavour, price and palatability. Recently, they have become concerned about the presence of 'additives', but nutritional value as such is normally taken for granted, and not seen as a cause for concern. There is a laudable pressure for more informative labelling of foodstuffs, but this does not seem to include the essential amino acids.

Where famine strikes, it is now generally held that malnutrition is due to a deficiency of total food intake, rather than a shortage of any one component of the

Table 4.1 Essential amino acid composition of seed protein. (Adapted from P. Payne (1983) *Seed Proteins*. Ed. J. Daussant, J. Mossé and J. Vaughan. Academic Press, London.)*

Amino acids	FAO pattern	Wheat	Rice	Maize	Soybean	Peanut	Broad bean
Isoleucine	278	253	290	225	319	224	333
Leucine	305	409	501	717	483	407	438
Lysine	279	174	239	169	429	218	476
Methionine + cystine	275	265	316	200	197	173	112
Phenylalanine + tyrosine	360	457	629	496	557	571	567
Threonine	180	192	235	225	269	171	284
Tryptophan	90	67	78	33	80	64	69
Valine	270	272	398	263	336	274	373

*Data expressed as mg amino acid/g nitrogen. Amino acids present at <90% of the FAO pattern have been underlined. The FAO pattern is that considered by the FAO as providing the ideal food protein for humans.

diet. It used to be thought that specific protein deficiencies existed in several parts of the world, but this is no longer seen as widespread. There may be a few pockets of protein deficiency, possibly associated with high cassava consumption, because cassava is particularly low in protein, but the idea that massive protein shortages exist in isolation is overthrown. Nevertheless, the daily bread still occupies a very significant place in our nutrition both in reality, and in how the general public perceives it. There may be ways in which biotechnology can impact on bread-making, but they will have to be pursued with great caution.

Baking

PROTEIN BODIES

Most of the protein in cereal seeds occurs in protein bodies, as it does in legumes, and apparently all seeds. Protein bodies are thus amongst the most important cytoplasmic structures known to food processors. It is interesting to note in passing that the only comparable structures found in animals are in the fat bodies of insect larvae, and also have a storage function. Only one such protein has been studied, calliphorin from the blow-fly, and it has some similarities with the seed storage proteins. As we have already pointed out in Chapter 1, the seed storage protein system is an attractive one as a site for expression of many transferred genes and may yet prove to be the optimum production system. It is possible that the insect larvae could also be used in this way. Protein bodies are roughly spherical, varying in size considerably from one species to another, and are contained within a single unit membrane consisting of phospholipids. This is not disrupted during hexane

Cereals: Baking and Brewing

extraction of neutral lipids, and is usually intact in 'defatted flours' of all kinds. When water is added to flour or any material containing protein bodies, it tends to enter the particle, which then swells, and eventually ruptures. During this period what is in effect a very concentrated solution of protein has an opportunity to form small gelatinous particles and this is the source of aggregates in some species. No mechanical process is effective in breaking up protein bodies. Their existence is sometimes taken advantage of in air-classification methods for enriching flours in protein. Selective flotation methods are also possible in principle but have not been developed to large-scale processes, despite several attempts.

There is almost certainly scope for the use of enzymes to assist in the disruption of protein bodies. The most obvious one is to use phospholipases to disrupt the membrane: it is unknown whether this is important in fermentations where the protein bodies are certainly eroded by enzymic attack (see Chapter 3) which is ascribed to proteases. There must also be some possibility of enzymatic attack during preliminary processing of the grain to make flour, though all the evidence is that, providing the water content is kept low, no enzymes are activated.

THE PROTEINS OF THE SEED

Protein body proteins are classified by a time-honoured method based on differential solubilities into, for cereals, glutenins, gliadins and prolamins. There are in addition an albumin fraction, which amounts to the very small cytoplasmic fraction of the seed. There is also a small amount of protein in the starch grains. Glutenins and gliadins together make up wheat gluten and occur in approximately equal amounts. The gliadins are the more soluble in dilute salt solutions, while the glutenins are the insoluble fraction. The other main group of cereal proteins are the prolamins, characterized by solubility in ethanol–water mixtures. The prime example of this is zein from maize, though both gliadins and glutenins do have some prolamin-like solubility. Although this nomenclature is still widely used (and will almost certainly go on being widely used in commerce) it is based on completely outdated analytical methods, going back nearly a century. The introduction of electrophoretic analysis has led to far more precise description of the cereal storage proteins. For a number of years after its introduction it was not possible to apply the technique because, like many other important food proteins such as actomyosin, the storage proteins could not be solubilized. Attempts to use solvents like concentrated urea solutions were not entirely successful, and have now been replaced by sodium dodecyl sulphate solutions (SDS), where necessary in combination with disulphide-breaking reagents. This is one more example of how the ability of SDS to act as a virtually universal solvent for proteins has dramatically increased our analytical capability.

The gliadin and glutenin fractions together contain about 70 different peptide chains. The soluble gliadin fraction has about 50, of molecular weight in the range 30–45 000, and with intra-chain disulphide links. It is now known that all the cereal proteins are closely related, and this size range shows many sequence homologies. They can usefully be classified into sulphur-rich and -poor groups. A remaining group, with molecular weights in the range 100–150 000 predominate in the glutenin fraction. In many cases the cysteine residues are located near the ends of the chains, and inter-chain disulphides are a feature of the glutenin fraction.

A number of attempts have been made to relate the presence or absence of subunit peptides to bread-making quality in different varieties, which have characteristic peptide patterns. These have met with limited success, though it is possible to follow a particular peptide through breeding programmes and produce varieties with almost any desired combination. The characteristic peptide patterns have been used in actions for breach of contract, to prove that the wheat supplied was not that specified.

The main contributor to the very distinctive elastic properties of dough is the glutenin, though satisfactory bread doughs certainly require the other proteins as well: they have been said to have a plasticizing role. There are theories which attempt to relate molecular properties to the viscosity, and the location of the cysteine residues mentioned above is thought to lead to very elongated structures. The subunits have a fairly random structure with numerous turns, which have been implicated in rubber-like elasticity properties. There is little direct evidence for the formation of extended chains. Indeed the connection between molecular structure and gross rheological properties is no better understood in this context than it is in the matter of gel structures in fermented soy products, or any other food where the structure is predominantly protein based.

DOUGH-MAKING

In dough-making, the flour and water are mixed to 'prove' the dough. This is a process essential to good baking quality and it leads to distinct alterations in the viscosity, which rises. Over-working leads on to a fall, and loss of quality again. The mixing is very vigorous: the energy input is quite enough to shear covalent bonds, and it is an expensive part of the process. Thus there have been many attempts to shorten the mixing period and reduce the energy input.

It now seems certain that the viscosity changes are due to disulphide bond rearrangements between the subunits. Addition of materials such as cysteine, sulphite or bromate, all of which are known to disrupt disulphide bonds (see also Chapter 3) have marked effects. The action of ascorbic acid has been investigated in detail since flour contains an ascorbic acid oxidase.

However, the most compelling evidence that disulphide rearrangements are involved comes from the role of glutathione. Good bread-making quality correlates with a low level of GSH, and all the other 'thiol reagents' are now thought to exert their influence, not directly on the disulphide bridges but by regulating the level of GSH. The enzyme disulphide isomerase (EC 5.3.4.1) has already been mentioned, and uses GSH as a catalyst in itself catalysing the re-shuffling of disulphides. It is not known whether this enzyme is present in wheat flour, though its presence would help to explain why GSH appears to be more important than other reagents. Attempts to add it, to see what it might do have only been reported once and with inconclusive results. The enzyme is not easy to obtain or assay and will probably have to be the subject of transgenic manufacture before proper trials can be carried out. It should also be borne in mind that only a

Cereals: Baking and Brewing

few out of numerous disulphides are important in proving, and disulphide isomerase might well catalyse unwanted ones.

EFFECT OF PROTEASES

When dealing with the effects of disulphide formation on the rheological properties of proteins, or even such basic matters as solubility, it is natural to think in the first instance of splitting the disulphides as a means of reversing the effects. Apart from disulphide isomerase, which is as yet untried, there are no enzymes to help, and it is not at all easy, chemically, to split disulphides (as opposed to rearranging them).

In fact reversal can often be achieved, in crude viscosity terms, which is what matters in food processes by splitting other bonds instead. This is certainly the case in dough-making. A large number of proteases have been found effective in modifying doughs for special purposes. Biscuit doughs use subtilisin (EC 3.4.21.14) and plant proteases such as papain (EC 3.4.22.2) can be used in any application needing weak doughs. It is possible to over-hydrolyse, and this is another example where it would be advantageous to have more precisely targeted proteases, especially now sequence data on the proteins are available. The enzymes are usually very crude preparations, and include amylases, and both endo- and exo-

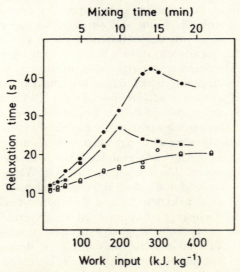

Fig. 4.2 Effect of adding lipoxygenase in the form of enzyme active soy flour on dough mechanical development: ■, without added enzyme; ●, with added enzyme; □, ○, mixed in nitrogen; ■, ●, mixed in air. The largest effect was obtained with both air and enzyme, and in the absence of oxygen there was no effect. Relaxation time is a rheological parameter, and work input is the total cumulative input in the time indicated. (From P.J. Frazier, F. Leigh-Dunmore, N.W. Daniels, P.W. Egitt and J. Coppock (1973) *J. Sci. Food Agric.* **24**, p. 421.)

peptidases. The latter produce some free amino acids which are important in flavour development.

OTHER ENZYMES

Lipoxygenase (EC 1.13.11.12) is routinely added to doughs, in the form of soybean meal, specially made so as to retain activity. It too can affect the redox status of the dough, and this is shown in Fig. 4.2 which serves to illustrate all these dough-modifying additives. It is however mainly used as a means of bleaching, because it removes the yellowish carotenoids, which the endogenous lipoxygenases do not attack.

STARCH GRAINS

Starch is actually about 70% of a typical flour and as such is bound to be involved in the structure of bread. As will be seen, an understanding of the role of starch has proved to be of great commercial significance. Flour contains α-amylases (EC 3.2.1.1) and β-amylases (EC 3.2.1.2) as well as an α-glucosidase (EC 3.2.1.3). Because of the well-known resistance of intact starch grains to attack, the extent to which the amylases degrade the starch depends on the extent of mechanical damage during milling. Poor quality grain, especially if it has been allowed to sprout, has much higher levels of amylase, which causes poor crumb structure. A shortage of amylase, or an absence of fractured starch grains, also causes poor crumb structure. Most significant, it leads to bread with a low moisture content, which stales rapidly. Staling is partly due to drying out and partly to retrogradation of starch grains, which depends on the presence of amylose.

It is now standard practice to add amylases. In the USA this is done by adding about 0.2% malted barley, while in Britain fungal amylases are used, said to be better because malt also contains potentially undesirable proteases and colours. This practice is partly because as dough mixing times are reduced by adding disulphide reagents, the time for endogenous amylase to act is insufficient, and the activity has to be increased. There is another reason. When Britain joined the EEC, a referendum was held and one of the arguments used was that Britain's bread was in peril. This was held to be because British bread manufacturers used a high proportion of hard wheats, mainly imported from North America. As a result it had an exceptional shelf-life, necessary because bread-making is highly centralized and needs an elaborate distribution system. Because of EEC agricultural policy it was felt that such wheats would have to be replaced by locally grown soft varieties. These were well known to produce bread which would go stale overnight, which fits fairly well with a continental European pattern of numerous local bakers, with fresh bread every day. The glutenin and gliadin levels of the two types of wheat are different—hard wheats have higher glutenins—and a number of projects to investigate the effect of added glutenin were undertaken.

Actually, the solution to staling lay in a different direction. During milling of hard wheats, the starch grains are extensively fractured, for mechanical reasons. Then, amylases digest rather more starch than they do in soft wheats, and increase the moisture retaining ability and decrease the amylose level. This is apparently

Cereals: Baking and Brewing

the reason for greater resistance to staling. By using added amylases it is possible to use soft wheat to make bread with long shelf-life. Problems have been experienced with bacterial amylases because they have too high a thermostability, but fungal amylases from *Aspergillus* are suitable, and are destroyed in baking. They apparently exert their main effects at 55–60 °C where setting, starch gelatinization and thermal inactivation all go on simultaneously.

A similar anti-staling effect has been claimed for a pentosanase (actually a cellulase with pentosanase activity) from *Trichoderma reesei*, though this has not been put to commercial use. Wheat contains 4–5% pentosans and rye even more.

British-produced wheats are, because of the wet climate, more likely to have excess amylase and it would be helpful to have varieties where the activity could be eliminated at relatively low temperatures. A modified gene might be the way to achieve this, but far more must be discovered about amylase structure before this could be done. It is notable how often in food applications the need is for a reduced thermal stability of the enzyme, in contrast to those wanted for purely chemical processes.

Flours are sometimes treated with chlorine which appears to increase the lipid-binding capacity of the flour, making it more suitable for cake formulations. Heating has similar effects. Electron micrographs of untreated starch granules show that amyloglucosidases cause extensive pitting of the surface in the course of digestion. Chlorine treatment eliminates this though it does not inhibit the enzyme. Removing surface protein by other methods also eliminates the pitting effect and it looks as if the effect of chlorine is somehow connected with the presence of surface protein on the granules.

British-produced soft wheats do contain less total protein content as well as less gluten, so some trials have been done to raise the protein content, but so far with inconclusive results.

LEAVENING WITH YEAST

After proving, the next stage in bread-making is the addition of yeast. When flour is mixed with water, the free water phase contains 10–15% dissolved solids, including glucose from amylase activity, most of which is small carbohydrates and this is what the yeast metabolizes. Baker's yeast can use glucose, fructose, mannose and galactose, and the disaccharides sucrose, maltose and trehalose but notably fails to utilize lactose and pentoses. In 'milk loaves' which contain added milk solids an *Aspergillus niger* galactosidase has been used to break up the lactose for this reason. It can also partly digest trisaccharides such as raffinose, leaving melibiose as an undigested residue. In aerobic conditions yeast can use a wide range of carbon sources, including ethanol. It has an absolute requirement for biotin, which it must obtain from the dough. Biotin is unique amongst the B vitamins in the very low levels in which it is needed in the diet so this effective depletion by yeast during leavening is not nutritionally significant. Some strains also need other B vitamins for optimal growth. The ethanol level in the free water can reach 4% during the raising process, and is presumably lost during baking. The carbon dioxide produced is of course responsible for the open sponge structure of the dough. Clearly the extent of this will depend on the amount of fermentable sugar, and the time the yeast is allowed to work. Additional fermentable sugar, such as sucrose, can be added at this stage if there is too little for some reason, perhaps because of insufficient amylose availability.

Taste panels can invariably detect differences in flavour between leavened and unleavened bread. It is possible to make bread, soda bread for example, where carbon dioxide is generated by other means to give a bread with a similar open structure, but lacking yeast. The aroma and flavour components are actually quite similar to those found in alcoholic drinks, and include a series of alcohols, notably pentan-2-ol and furfural which have a quite characteristic fresh bread aroma. As seems inevitable with large-scale processing, efforts are continually being made to reduce the time needed for this stage in the process. This can be done by controlling the temperature, and by adding fermentable sugars. But this has consequences for flavour development, even to the extent that additional flavours may have to be added just before baking.

There is a market for frozen doughs, and they are also sometimes used for production convenience. Certain types of pastry are best made with doughs allowed to develop slowly at 5–6 °C for long periods. This is more to do with controlling the amylase activity than yeast.

Among individual enzymes, lipases are present in dough, and can also be produced by yeasts. It is believed that minimal levels are best: fungal lipases released as a result of contamination lead to rancidity. This is doubtless due to the release of free fatty acids, followed by their oxidation. Phytases are important. They are present and active in flour, and normally hydrolyse all the phytic acid to inositol and phosphate. In whole meal up to half the phytic acid can survive to the loaf, and yeast also plays a part in phytase activity. Unleavened wholemeal bread is said to have higher levels of phytic acid, which can lead to difficulty in absorbing calcium by its consumers.

CONCLUSIONS

The final stage in baking, heating the shaped dough, stops all enzyme activity. The overall impression is that while enzymes are widely used in baking, and some of them have become highly significant in commercial terms, in allowing a wider choice of raw materials, genetic manipulation has so far had no impact. It may find its first application in methods for the production of baker's yeast, itself a significant enterprise. There is certainly scope for improved proteases and amylases, in the form in which they are actually used. This is more to do with uniformity and shelf-life, i.e. consistency and storage properties, than the precise mode of action. Indeed while the precise action of proteases on dough remains unknown it is difficult to see how better ones could be specified. There are trypsin and papain inhibitors in flour which could be eliminated by suitable breeding, though they are of little significance.

There is always the possibility that novel enzymes, like disulphide isomerase or pentosanases may become available. Pentosanases may be the best target, and could be used to modify dough properties, and would also be very useful in legume processing.

Brewing Beer and Vinegar

BREWING

Beer is made mostly from barley (*Hordeum* spp.) by degradation of the starch to maltose followed by a yeast fermentation to produce ethanol. Two further significant branches of the food industry are based on beer: oxidation of the ethanol to make acetic acid, and hence vinegar, and its distillation to produce whisky. (Strictly, in English usage, beer must contain hops and is otherwise known as 'ale' but this discrimination has almost disappeared from common use.)

In the UK, about 5 million tonnes of barley is fed to animals every year and about 2 million tonnes is used for malting, much of which is used for whisky. Several million tonnes are exported, mostly destined for beer making in continental Europe. Production of malt in the USA is about 2 million tonnes annually, almost all used for beer. Beer is mostly made to be drunk, and dozens of variations, in flavour and ethanol content, stored in barrels, bottles or cans, drunk at room temperature or chilled, are made.

All beers ultimately depend on the starch granules of the barley, making up some 85% of its weight. Degradation of starch has already been mentioned in Chapter 2 in connection with glucose manufacture, and above in the context of bread-making. The distinctive feature of brewing is that as much as possible of the starch is degraded by the enzymes of the barley itself, which are brought into action by germinating the barley and then stopped completely by heating before adding yeast. In baking and in some whisky production the enzymes are not stopped before adding the yeast and degradation of starch and fermentation proceed simultaneously.

Starch Grains

The mechanisms by which starch is synthesized *in vivo* are well delineated though as usual little is known about the enzymes responsible. They contain the synthases which make $(1 \to 4)$ links and leading to amylose, and the branching enzymes which make $(1 \to 6)$ links and lead on to amylopectin. It appears that some of them are membrane bound while others are not, and there is a possibility that the membrane-bound ones are primarily responsible for amylose. The enzymes are present in multiple forms, but most of the work has been done on maize, and does not necessarily apply to barley. One variety of barley, high in amylose, is believed to lack one of three identifiable branching enzymes. Conversely, an increase in branching enzyme activity leads to phytoglycogen, which is more extensively branched than amylopectin and is water soluble. The thrust of this work is to find ways of controlling the proportion of amylose and amylopectin, and is likely to find more direct application in high-fructose syrup manufacture and the starch-based chemical industry, than in brewing.

The starch grain itself is highly structured. Figure 4.3 gives a current view. It is important because in this application, unlike others, it is the normal mechanisms for starch mobilization that are used, at least in the initial stages of digestion, and this implies an attack on the intact *in vivo* grain.

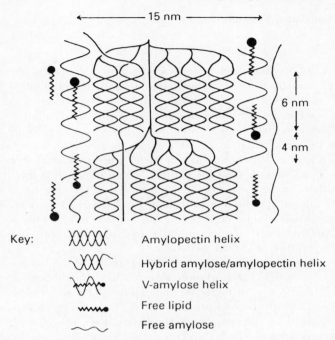

Fig. 4.3 Diagrammatic representation of the structure of a starch grain. There is, in addition, small amounts of protein in pockets on the surface of the grain. (From J. Blanshard (1986), *Chemistry and Physics of Baking*. Eds J. Blanshard, P.J. Frazier and T. Galliard. Royal Society of Chemistry, London.)

Barley contains two groups of starch grains, the most numerous with 5 μm diameter and another, making up 90% of the weight, but much less numerous, with an average diameter of 25 μm. They differ slightly in composition, the larger ones containing more amylopectin, while the smaller have more protein, and gelatinize at a slightly higher (54 °C) temperature than the main ones (52 °C). A variety of barley exists which has only amylopectin, and there is also one with 43% amylose, as opposed to the more usual 20%. Most work on mutation in starch synthesis has been done on maize, and so far the techniques of gene transfer have not been used. Clearly it would be possible to provide sources of barley starch with some control over the amylose:amylopectin ratio. This is more likely to be developed for the chemical industry based on maize and potato starch than in the food aspects.

Little is known about the deposition of starch to form the grains. Since the starch grain is highly structured, there is presumably a spatially organized synthetic mechanism and some means of providing amylose and amylopectin in the required proportions. Other experience suggests that this most probably involves localization of enzymes on membranes, consistent with the observation that at least some of the enzymes are when isolated, membrane bound.

There has been some speculation that the centre of the grain may contain a 'nucleation point' of glycoproteins. A similar idea was associated with protein bodies, though it is not now accepted, since no such nucleus could be found. It would be interesting if the synthetic mechanism could be altered to produce a relatively disordered grain, since this would usefully affect the way in which degrading enzymes attack.

Amylase Inhibitors
Amylase inhibitors have been found in wheat and rye, and seem to resemble the protease inhibitors in some respects. They form stable 1:1 complexes with the enzymes, are often proteins or peptides, but can be oligosaccharides and are widespread in microorganisms. Some of the fungal protein ones are heat stable. Their function is quite unknown, though they might serve to control amylase activity *in vivo*. It is possible that these inhibitors play some part in the malting process, but the matter seems not to have been investigated. Comments already made about the potential utility of designed proteases with high specificity towards protease inhibitors apply equally to amylase inhibitors.

Attempts have been made to use phaseolamin, the inhibitor from *Phaseolus vulgaris*, as an inhibitor of mammalian digestive amylase. More specifically, they were proposed as a dieting aid for obese people, but led to flatulence. This is a common result of the presence of large amounts of undigested carbohydrate in the human gut. Similar results occur with the eating of untreated legume carbohydrates. This use has been prohibited by the US Food and Drug Administration.

Starch Grain Degradation: Malting
In brewing this takes place in two steps. On malting floors, barley is dampened to start germination at about 40% moisture and then kept for 3–6 days, depending on temperature. Amylase activity develops quickly, and includes both the α- (EC 3.2.1.1) and β(1→4) amylases (EC 3.2.1.2) and some (1→6) pullulanase (EC 3.2.1.41). The malt is then air-dried with some care to retain most of the activity, though it can be used directly without first drying, when it tends to have a somewhat higher activity. The next step is to extract the ground malt with warm water, around 60 °C. The enzymes break the starch down further. Some typical compositions are shown in Table 4.2. The temperature is high enough to give some gelatinization of the starch which undoubtedly aids the breakdown. Little is known of the way in which amylases interact with grains during normal mobilization. A recent study of the way in which β-amylase interacts with starch gels, which used a new fluorescence method to follow the mobility of the enzyme, offers some hope of useful insight into the mechanism.

In this technique, called 'fluorescence recovery after photobleaching', or FRAP, the enzyme is labelled with a fluorescent group, usually fluorescein. By first bleaching a small area with an intense irradiation from a suitable laser, followed by a lower level of irradiation to excite fluorescence, it is possible to observe the reappearance of fluorescent groups in the bleached spot, due to diffusion from the surrounding area. An estimate of diffusion rate can be obtained which gives useful insight into the nature of the surface interactions. The method was first introduced to study the fluidity of membranes, but has been extended to the study of proteins on surfaces. There are numerous related problems in food processing—the same

Table 4.2 Composition of mashed wort at three temperatures

Component (g sugar/100 ml)	Temperature* (°C)		
	62.2	65.5	68.8
Glucose + fructose	1.12	0.98	0.81
Sucrose	0.40	0.40	0.45
Maltose	4.30	4.19	3.92
Maltotriose	1.49	1.55	1.63
Dextrins	2.03	2.24	2.52
Total	9.34	9.36	9.33

*The temperature is that at which enzyme digestion of the starch took place. The dextrin level is very temperature-sensitive, and affects the character of the beer. (Data from I. MacWilliam (1968) *J. Inst. Brew.* **65**, pp. 38–54.)

method has been used to follow lipase mobility (Chapter 5)—while protein bodies await study. In this kind of interaction, surface diffusion is important, and is modified by specific interactions. It was found that β-amylase mobility through the gel decreased sharply when it was inactivated with iodoacetamide. This phenomenon, if general, must offer some scope for enzyme modification techniques. Figure 4.4 illustrates some results with maize starch that emphasize how complex the attack must be in ordered heterogeneous systems.

It is possible at this point in the process to add both extra substrates in the form of non-barley starches, and extra enzymes. At least 60% of the storage proteins are broken down, and provide some of the nitrogen needed by the yeast. If large amounts of pure starch are added it will be necessary to add further nitrogen sources. It is not clear to what extent this is actually done in normal brewing practice. In some countries it would certainly be prohibited. Glucans from cell walls also become extractable.

The 'dextrins' referred to in Table 4.2 is the limit dextrin left because the amylases cannot break $(1\rightarrow6)$ links or the $(1\rightarrow4)$ links near them in amylopectin. The level is related to the quality of beer. Light beers have a lower level, and it can be adjusted with added glucoamylase, which breaks all these links. After standing for the appropriate time, the wort is boiled which effectively stops all enzyme activity, and also sterilizes it. After cooling yeast is added, and fermentation proceeds until all available carbohydrate is consumed. Both yeast and carbon dioxide are byproducts of brewing.

One large brewer is currently investigating the use of yeasts genetically modified to produce proteins of pharmaceutical interest. So far a yeast expressing human serum albumin has been made. The idea is that after the normal brewing, the yeast will be switched, by a pH or sodium chloride alteration, to secrete the protein. It remains to be seen whether this can be successfully linked to beer manufacture. Pharmaceutical manufacture requires much higher standards of sterility than are usually found in breweries.

The use of papain to remove chill haze from beer has already been mentioned in

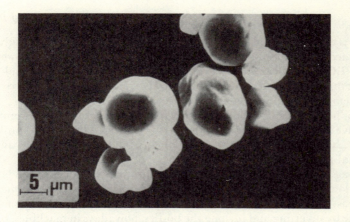

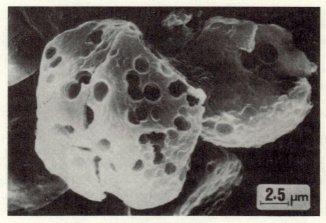

Fig. 4.4 (Top) intact maize starch grains, seen in the scanning electron microscope. (Bottom) the same grains after 60% digestion by pancreatin, a mixture of rat pancreatic enzymes. Degradation is clearly heterogeneous. (From H. Fuwa, T. Takaya and Y. Sugimoto (1980), *Mechanism of Saccharide Polymerisation and Depolymerisation*. Ed. J. Marshall. Academic Press, London.)

Chapter 3. It would be ingenious if the yeast could be made to release a papain-like enzyme at the appropriate time. The foam stability of beer which is an important consumer attribute, depends on the presence of small amounts of surface-active protein, and it is important that proteases which eliminate chill haze do not at the same time remove these proteins. Papain is satisfactory but no-one knows why. These proteins too must be candidates for expression in yeast. A Japanese brewer has claimed to have inserted a gene for an amyloglucosidase into yeast, which is able to directly ferment starch, without prior malting.

Beer is usually pasteurized as a final step before storage. There are still a few beers sold live, with viable yeast in them.

One of the changes which occurs in beer maturation is the oxidation of α-acetolactic acid into diacetyl, and then acetoin, by yeast reductases. In many cases, this is the rate-limiting step, and as always there are commercial advantages to be gained from speeding up maturation. An enzyme, acetolactate decarboxylase (EC 4.1.1.5) can convert acetolactase directly to acetoin, and it seems likely that this will be used in the near future as an aid in brewing. At present there is a problem with finding a good source for a usable enzyme.

Hops, or hop extract, is added to the wort. It was probably added originally as a preservative but it is now integral to the flavour of beer. There would probably be interest in alternative sources of hop components, but this is many years away from commercial practice.

Champagne and sparkling-wine manufacture depends on a secondary yeast fermentation in the bottle to produce carbon dioxide. This of course means that there is yeast in the bottle that has to be removed. In the true champagne method this is done by inverting and turning the bottles, nowadays in machines, but this is still a lengthy and expensive operation. When all the yeast has collected on the cork, it is removed, often after freezing a plug in the neck, and the bottle re-corked. Some manufacturers are now testing the use of encapsulated yeast, which is much easier to remove. Something similar could probably be used in beer manufacture, although while there are continuous brewing processes in operation, so far none of them use immobilized yeast.

VINEGAR MANUFACTURE

Vinegar, is made from cider or wine, and that based on beer is distinguished as malt vinegar. It is used in large amounts in Britain and Northern Europe, and the USA in the manufacture of a great variety of pickled products. Certain parts of England, notably the West Midlands show a strong preference for acidic foods, and consume large quantities of pickles and articles like Worcester sauce. Lactic fermentation as a means of preserving meat, common in southern Europe—the preservation is due to a combination of low pH and low water content—was never developed in England, perhaps because of a ready supply of strong vinegar.

An enormous variety of fermentable substrates, ranging from coconut extract to peach juice have also been used to make veinegar. However in practice most is made from beer or wine. In most countries, the most convenient local alcohol source is used, including ethanol from petrochemicals. Vinegar in the UK is made from beer by an aerobic oxidation, traditionally carried out by *Acetobacter* species (formerly called *Mycoderma acetii*), grown on birch twigs. Biotechnology has had no impact so far. Malt vinegar tends to be colourless and is turned brown by the addition of caramel.

CONCLUSIONS

It is difficult to believe that such a large and varied industry as brewing will not find applications for biotechnology. Enzymes are used, and have been for some years, but there is no sign of any impact from genetic engineering techniques, other than a modified yeast to make what amounts to new byproducts. Perhaps the very lengthy period of empirical experiment, together with a tendency over the last century to do a good deal more research and development than the food industry in general, has left fewer opportunities for novel enzyme applications.

Chapter 5

Lipases and the Minor Components, Emulsifiers, Stabilizers and Flavours

Introduction

A typical ingredient list, taken from the label on a bottle reads as follows:
- sugar
- dried skimmed milk
- dried whey
- milk-chocolate
- cocoa
- vegetable fat
- dried glucose syrup
- flavourings
- stabilizers
- potassium bicarbonate
- sodium bicarbonate
- edible acid casein
- emulsifier–glycerol monostearate

They appear in order of decreasing amount by weight: the label happens to apply to a chocolate drink, but is typical of dozens of possible examples.

The major components, those appearing first, have largely been the subject of Chapters 1–4. The minor components, those present in much smaller amounts—emulsifiers, stabilizers and flavourings—are the subject of this chapter.

Lipases are the enzymes most likely to be used in flavour and emulsifier work, and have also as it happens been used in a new potentially large-scale application in cocoa butter preparation.

Properties of Lipases

Glycerol ester hydrolases (EC 3.1.1.3) are universally present, as are their substrates the triacylglycerols. They are digestive enzymes, whether they are secreted by fungi, bacteria or the human pancreas, and are not involved in anabolic processes. They possess, uniquely, the ability to act at an interface between water and a non-aqueous phase, which may be a drop of triglyceride itself or a solution of it in a paraffin such as hexane. This is the feature which distinguishes them from esterases, which can split similar bonds in water-soluble esters, for example triacetylglycerol, or tributyrin. While lipases can usually split soluble esters, the rates are low compared with those obtained in the interfacial situation.

Lipase activity depends on the available surface area, and is usually measured on emulsions with a very large excess of surface compared with the amount of enzyme protein present. It is not possible to interpret rate measurements in the same way as for enzymes acting on soluble substrates, which has led to rather restricted or even erroneous data from classical enzymology. For example, substrate specificity cannot be interpreted very easily when the factor which distinguishes one triglyceride from another may be due to its ability to emulsify, rather than the way in which the constituent fatty acids interact with the enzyme active centre. Since monoglycerides, one of the products of the reaction are also emulsifiers, the rate of hydrolysis can suddenly accelerate simply because a greater surface area has become available.

It used to be said that lipases are hydrophobic enzymes because they do interact strongly with lipids. An old preparative method involved adsorption onto waxes. This is wrong. Calculation of the hydrophobicity coefficient (see Chapter 3) from the amino acid composition shows that many of them fall into the hydrophilic end of the overall distribution for globular proteins. On top of this they are frequently glycoproteins. They contain $\leq 10\%$ carbohydrate. In the case of *Aspergillus niger*, galactose, N-acetylglucosamine and mannose residues are in the ratio 1:1:4. There are several glycosylation points—TGN—in the known amino acid sequences, and the carbohydrate probably occurs in several distinct regions. The detailed structure of the porcine pancreatic lipase glycan has recently been determined and contains, in addition fucose, but is present in a single-branched chain. It is thought that the chain varies between different lipase molecules.

The function of the carbohydrate is unknown. Partial removal with mannosidases has no effect at all on the activity. It used to be thought that glycosylation in some way marked proteins for secretion, but the evidence for this is now weak, and lipases are known to be synthesized in a pre-form, with scission on membrane transit. The most plausible advantage attached to glycosylation is protection against protease attack. Secreted enzymes, like fungal lipases are accompanied by potent mixtures of proteases as part of the digestive battery. In one case investigated, *A. niger* lipase was untouched by high levels of trypsin or chymotrypsin. In other examples, a series of proteins with lipase activity has been obtained from culture media, with every indication that they are proteolytic fragments of one secreted lipase. The matter remains unresolved, but is not unimportant because current attempts to make lipases by expression in *E. coli* will result in unglycosylated molecules. However, while human and porcine lipase are glycosylated, equine, ovine and bovine are not, and some bacterial enzymes are also probably carbohydrate-free.

There has been much interest in lipases in the last few years, because they are seen as useful for addition to detergents. In this work, alkali-stable lipases are required, and have been found in *Pseudomonas* species. These are however unacceptable for commercial use, so the gene has been transferred to *Aspergillus* and *Bacillus subtilis*, which are the source of detergent proteases and well-known already. Food applications have also been developed, while the ability of lipases to work on materials soluble in non-aqueous solvents, and exert chiral specificity has led to much interest in the fine organic chemicals field, including some possible flavour component syntheses.

High-resolution crystallographic structures for lipases have not yet been published, but low-resolution ones, for *Geotrichum candidum* lipase, and for equine lipase are available. It suggests that the active centre is at the bottom of a cleft, which may explain the varying specificity of the enzyme. As is suggested in Fig. 5.1, depending on the stereochemistry, either the 1- or the 3-position can reach the right point, but the 2-position might not be able to. Most lipases appear to attack the 1 or 3 links, but some can hydrolyse all three. The question is obscured because acyl groups can migrate between the three positions spontaneously, so that complete hydrolysis can be obtained with both kinds of lipase. There has been one report that a lipase from *Geotrichum* preferentially releases Δ^9 unsaturated fatty acids. The active centre contains a serine, which is acylated during the reaction and there is some homology with serine proteases. Protein engineering methods have been employed on a cutinase from *Pseudomonas putida*, which has high lipase activity with a pH optimum in the range 7–11. The active site sequence is,

$$-Ser-Gly-His-Ser-Gln-Gly-Gly-Gly-Ser-$$
$$126$$

and when Ser-126 was replaced by an alanine residue all activity was lost. Serine esterases of which lipases are examples always have a His–Ser–Asp catalytic triad. Systematic replacement of all six His residues, one by one, with a Gln residue, resulted in loss of activity only when His-206 was replaced. An attempt to find out which of the 12 Asp is the essential one, by a similar replacement method, is at present in progress. Many other alterations can be made. Replacing the Gln-127 by a threonine is claimed to lead to a change in the rate of inter-esterification as compared with hydrolysis.

Although a long way from commercial application this is a good example of how protein engineering is beginning to offer possibilities of altering enzyme activity in a useful way. There would obviously be considerable interest in making lipases with fatty acid specificity.

There is almost certainly a hydrophobic patch on the enzyme surface which is able to bind strongly to the lipid surface. Proteins with hydrophobic patches, in solution, would be expected to dimerise by their interaction, and some lipases may do this. Non-associating systems show a steady decrease of the sedimentation

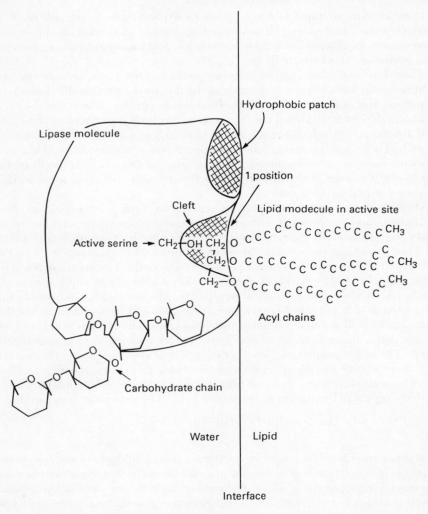

Fig. 5.1 The main features of the lipase molecule. Lipase action on triacylglycerols: approximately to scale, is shown how a cleft locality for the active serine could lead to (1→3) specificity. The substrate cannot be inserted so that the 2-position reaches the active serine. A hydrophobic substrate binding patch is also indicated, and a carbohydrate chain which some lipases have.

coefficient (S) as the concentration increases: in crude terms, as the viscosity increases S ought to decrease. For some lipases, such as *Geotrichum candidum* there is a report that S increases with concentration, presumably because of an association process. Others, such as *Chromobacter* and *Pseudomonas* lipases show no such behaviour, which is actually evidence that they do not have significantly exposed hydrophobic patches in solution (N. Simkin, S. Harding and M. Tombs, unpublished). This behaviour may explain some otherwise puzzling observations with inhibitors. Lipases are inhibited by isopropyl-fluorophosphate and similar serine esterase inhibitors, but only in the presence of an interface. Possibly dimerisation, or the structure of the enzyme itself, so obstructs the active centre that the inhibitor can only gain access when the monomer is absorbed to a surface, which exposes both the active centre and a binding site as a result of conformational change.

Vertebrates use bile salts as lipid emulsifiers in the gut, and of course all surfactants necessarily fill the interface and thus potentially obstruct the access of lipases. In mammals, and some fish, it is known that another peptide is produced by the pancreas, which is able to assist the lipase to penetrate the surface layers. Called 'co-lipase', it apparently has no counterpart in fungal or bacterial systems. It might be part of the ancestral lipase molecule which has become detached during processing in the pancreas, and analysis of the amino acid sequences for lipases which are just beginning to appear will probably reveal which part of the fungal lipase corresponds in function to the co-lipase. Nevertheless, it is always the case that the lipase will have to compete for its place in the interface with other surfactants, and it is unknown at present whether it is particularly well adapted for this or not. One study, which used fluorescent labels on lipases to determine their surface occupancy showed that they could easily be displaced from heptane surfaces by other proteins, but when triglycerides were present, they adhered more strongly. The interfacial location of lipases dominates attempts to use them in processes.

INTER-ESTERIFICATION

The most significant food process using lipases will shortly be in large-scale operation, and involves inter-esterification to make cocoa butter substitutes. Chocolate softens at 35 °C and melts at 37 °C, a property which is conferred by the presence of stearyl, oleoyl, stearyl glycerol (abbreviated SOS) and palmitoyl, oleoyl stearoyl glycerol (POS) in the fat. These triglycerides are predominant in cocoa butter, which is expensive. A process (Fig. 5.2) has been known for some time, in which mixing stearic acid, which is cheap, with tripalmitoyl glycerol, which is also very abundant (palm oil mid-fraction), and using chemical catalysis, leads to a shuffling of the fatty acids, and the formation of at least some of the valuable POS and SOS. These can be separated by conventional triglyceride fractionation. After inter-esterification and crystallization of PPP and SSS from the mixture the remainder is remarkably close to cocoa butter in composition. An alternative triacylglycerol which could be used is the high oleate sunflower oil. However, side reactions in chemical inter-esterification—which can be avoided by the much milder conditions of enzymic catalysis, as well as the possibility of making use of $(1 \to 3)$ specificity, and the general climate of opinion amongst regulatory agencies—made an enzyme-based process attractive.

In preparing lipase for use in the process, a very crude lipase preparation, where only about 1% of the protein was actually the active enzyme, from *M. miehei*, was

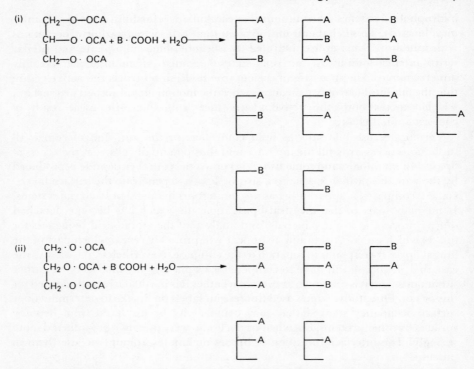

Fig. 5.2 The inter-esterification reaction. A and B represent acyl chains. (i) The products obtained if an enzyme able to attack all three positions in the triacylglycerol is used. (ii) The much smaller number of products obtained when an enzyme with (1→3) specificity is used. Diglycerides and monoglycerides appear in small amounts in the inter-esterification reaction because some water is always present. Since the triacylglycerol can contain three different acyl chains, this is the simplest possible case, and the reduction in complexity is greater when the three acyl chains vary.

precipitated onto a support with acetone. After drying, the powder is packed into conventional columns, and the reagents pumped through, dissolved in heptane. A small amount of water is essential and appears to mobilize the lipase into the interface and start the inter-esterification. Very careful control of water activity is obviously needed, since hydrolysis must be avoided as much as possible. In trials, about 14% of the triglyceride was brought down to diglyceride or monoglyceride. The lipase activity was reasonably stable, but problems were found of poisoning by some feedstocks, and preliminary purifications may be needed. The rate of inter-esterification is sufficient to permit operation on a steady-state flow-through basis. The columns are run at 40 °C with a 10-min dwell time. The plant has been run without interruption for several months, and many tonnes of lipid can be processed for each kilogram of prepared enzyme support. A number of different crude supported lipases function perfectly well, and the nature of the support and its flow properties are more significant in the successful operation of the plant. A further development, in which thermostable lipases are used on a substrate of molten fats, avoiding the use of hydrocarbon solvents, might be achieved. One possibility is the use of methyl esters of fatty acids, in this process methyl stearate which has a low melting point. There are also considerable possibilities in the use of fatty acid specific lipases. These are not available at present but it is reasonable to expect that protein engineering applied to lipases will lead to their development.

OTHER APPLICATIONS OF LIPASES

Lipases can catalyse the following types of reaction, in addition to simple hydrolysis and inter-esterification, at rates likely to be commercially useful.

- *Alcoholysis*, e.g. methyl palmitate + octanol→octyl palmitate + methanol
- *Acidolysis*, e.g. triolein + lauric acid→tridodecyl glycerol + oleic acid
- *Esterification*, e.g. oleic acid + octanol→octyl oleate + water
- *Aminolysis*, e.g. heptylamine + methyl palmitate→heptyl palmitate + methylamine

In addition, especially if vacuum distillation is used to remove water, they can be used to make wax esters, and many other similar esters, including galactose esters of fatty acids.

There are no large-scale applications of lipases other than inter-estification, and a possible future application in fatty acid residue randomization, though they have found some use in the generation of flavour components. In most cheeses, the flavour depends on the presence of shorter-chain fatty acids. Typical Swiss hard cheese contains propionic acid, which arises from lactose fermentation by *Propionibacterium*, while some US cheeses such as 'Romano' contain butyric acid as a major flavour component. Cream, which has been partially hydrolysed by added lipases, has been used as a source of short-chain fatty acids, for addition to cheeses. Also, crude lipases are sometimes added to cheeses to intensify flavour, by the same route.

Milk itself contains a lipase, identical to the lipoprotein lipase present in blood. It also contains an 'activator' peptide, which facilitates its attack on milk lipids, and appears to be similar to the colipase of the gut. It has been implicated in the development of fruity off-flavours in milk products, but also plays a part in flavour development in cheese and yoghurt. However, the main interest for the future employment of lipases lies in their use for synthesis of esters. The ethyl esters of short-chain fatty acids, up to C_8, caproic, are important flavour components. In the relative absence of water, lipases are capable of synthesizing esters by reversal of their usual hydrolytic reaction. The enzyme has a minimal requirement for water to maintain its structure but this appears to be little more than a partial hydration shell. Provided that there is a roughly equal weight of water and enzyme, the other problem is that the build up of glycerol may damage the enzyme. In fact, lipases can withstand at least 80% glycerol at temperatures up to 50 °C, though 90% does lead to conformational loss and activity loss. In some processes the products of reaction, such as methanol will be much less well tolerated. The problem is to bring the lipase into contact with the substrates dissolved in a non-aqueous solvent, in a suitably efficient manner. Two approaches have been investigated. One uses the idea of reverse micelles to disperse the lipase through the solvent. Reverse micelles are essentially a single polymer molecule, with little more than a single-layer hydration shell, surrounded by a shell of surfactant. By far the most effective is bis-2-ethylhexylsulphosuccinate (AOT) a cationic detergent, though lecithin has also been used. Lipases dispersed in this way do have some activity against substrates dissolved in the organic phase. The activity is in some ways surprising since the substrates must penetrate the surfactant layer: there is also a danger of encapsulating an inactive dimer form of the enzyme. It is also obviously desirable to use a purified enzyme, rather than the usual crude mixtures. Purified lipases are just beginning to become commercially available, since they are required for some diagnostic tests of the level of serum lipoprotein, but they have not as yet found any food use. Lipases are relatively easy to isolate by using their special ability to interact with alkyl chains on columns such as octyl dextrans or with lectins for the carbohydrate containing ones. A more difficult problem is finding a microbial source which does not produce nicked enzyme.

A rival approach uses derivatives of the enzyme itself to obtain dispersal through the organic phase. It has been shown that lipases with polyethyleneglycol chains covalently linked, disperse readily in hydrocarbons, and are active. The problem here is, apart from the need for purified enzymes, that it will not be possible to obtain active derivatives of all of them. The number of available lysines in the surface is limited—in the successful example, four were available, but in others only one reacts. It is possible that carbohydrate-free versions of lipases, made in bacterial hosts, will be superior in these applications, and looking further ahead the insertion of some well-placed lysine by protein engineering should also be possible. Solubilization in organic solvents by attachment of polyethyleneglycol has also been effective for chymotrypsin, catalase and horseradish peroxidase, and the system as a whole appears to be much more stable than reverse micelles.

Another approach is to localize the lipase in membranes, which can be made hydrophobic, so that an interface can form in the membrane itself. Such systems have been investigated for the synthesis of sucrose fatty acid esters, potentially for use as emulsifiers. There are conflicting claims for the effectiveness of this kind of arrangement.

Serious use of isolated lipases in the food industry is still some way off, even for relatively high-value flavour components. They will probably be used in chiral syntheses first, for pharmaceuticals. It has recently been found that dry lipases dispersed in hydrocarbons remain active at 100 °C for long periods. This is an interesting observation, though more likely to be useful in organic synthesis than in food applications.

Emulsifiers

Emulsions are the fundamental structure of a number of foodstuffs. The most familiar are milk, cream, ice-cream, mayonnaise and similar dressings, and chocolate, which are emulsions of lipid in water, and butter and margarine which are dispersions of water droplets in a lipid continuous phase.

In food manufacture, there are two kinds of added components: emulsifiers which help to produce the dispersion of lipids, and surfactants and stabilizers which help to maintain the emulsions over the required timescale. Stabilizers have other functions as well, in giving the necessary viscosity to food products, and will be discussed below. Food emulsifiers are typically molecules with hydrophobic regions, with an affinity for neutral fat at one end, and hydrophilic groups with an affinity for water at the other. The role of proteins and peptides as emulsifiers, which can meet these structural requirements, and their production with the aid of proteases has already been treated in Chapter 3.

PHOSPHOLIPIDS AND PHOSPHOLIPASES

Phospholipids are important emulsifiers, and are present in many foods. Mayonnaise and the hollandaise type of sauces depend on the phospholipids and lipoproteins of egg-yolk to form the emulsion. In milk, nearly all the phospholipid is found in the membrane round the fat globules, along with glycoproteins and some sterols. However the major phospholipid used in food processing is lecithin, phosphatidylcholine, which is obtained as a so-called 'gum' during the refining of soy-bean oils. The structure of phosphatidylcholine, and the point of attack of various phospholipases is shown in Fig. 5.3. Phospholipases for laboratory work are typically obtained from snake venom, and have not so far been produced in a

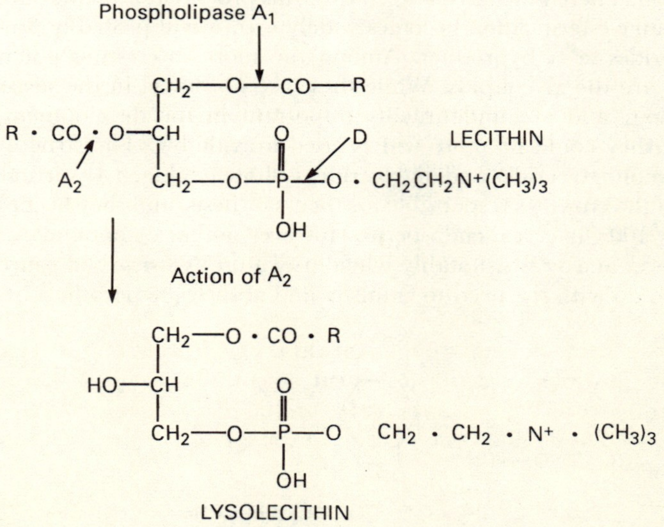

Fig. 5.3 Lecithin and the points of action of phospholipases A1 (EC 3.1.1.31), A2 (EC 3.1.1.4), C (EC 3.1.4.3) and D (EC 3.1.4.4). R is a fatty acid residue, usually C_{16} or C_{18}.

more convenient host by gene transfer. This might be difficult, since phospholipases could be highly damaging to the organism—they are not present in snake venom by accident—but almost certainly an inactive pre-form is made by the snake itself and might be transferable.

A great deal is known about phospholipases, including the crystallographic structure of A2. It is interesting to note that they belong to a completely different class of enzyme from the lipases, and are not serine hydrolases. The mechanism appears not to involve an acylation step.

The extent to which phospholipases can attack neutral triglycerides is unclear, though they probably do have some activity. Modification by mutagenesis might well lead to a novel series of lipases. In addition there is a role for phospholipases in disrupting the membrane round protein bodies as an aid to protein extraction from seeds. Lysolecithin is now used in some specialized emulsifier applications, and is for the present made by chemical hydrolysis. Lysolecithins are the main lipids present in cereal starch grains, and are believed to interact in specific ways with the amylose, and to have some effect in starch degradation.

MONOGLYCERIDES AND GLYCOLIPIDS

Monoglycerides are much more surface active than either diglycerides or triglycerides, and are manufactured for use as emulsifiers. Indeed in our example label, glycerol monostearate was the one used, and would have been made by hydrolysis of tristearin. While they can be made by using lipases to hydrolyse triacylglycerols, at present they are made by chemical hydrolysis followed by fractionation. There is no reason to change this process at present, but if enzyme-catalysed inter-esterification becomes widely used it will probably produce some monoglycerides as a byproduct. Among the more interesting potential novel emulsifiers are the glycolipids. While these are not novel in the sense that they occur in flour and are undoubtedly important in the development of dough properties, they could be more widely used if available. The structure of some galactosyl monoglycerides and diglycerides is shown in Fig. 5.4. Virtually nothing is known of the enzymes responsible for their synthesis, and they occur at levels of about 0.3 g/100 g in cereal endosperm. However similar compounds are made by some bacteria and yeasts, notably when grown on hydrocarbon sources. This is probably to do with the need to emulsify and absorb the paraffins. Studies have

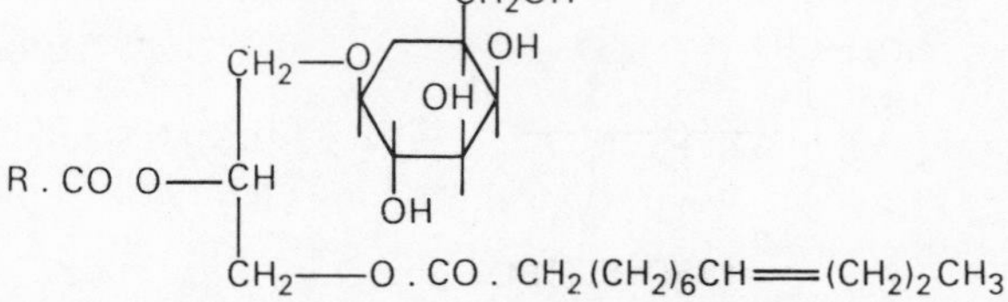

$$\text{R.COO}-\overset{\displaystyle CH_2-O}{\underset{\displaystyle CH_2-O.CO.CH_2(CH_2)_6CH=(CH_2)_2CH_3}{\overset{\displaystyle |}{CH}}} \begin{array}{c} CH_2OH \\ OH \\ OH \\ OH \end{array}$$

Fig. 5.4 A monogalactosyl glycolipid. The digalactosyl derivative is also common. R represents a fatty acid which may be either saturated or unsaturated, and is usually long chain, C_{16} or C_{18}.

been made on *B. subtilis* with a view to optimizing production, but any food-use application must be many years away. It is quite likely that a plant source, modified to make larger amounts, would be commercially superior. Another possible route would be the *in vitro* synthesis of similar, but not necessarily identical, galactosyl lipids, from galactose and fatty acids by using the isolated enzymes. Unfortunately, such a synthesis is likely to involve ADP or UDP, and it would be necessary to search for enzymes without these requirements. Such a quest was successful in the case of fructose isomerase, where the same problem was faced. Chemical synthesis would be very difficult and prohibitively expensive. The total market for emulsifiers is not large, and there is no pressing need for new ones, except as cheaper alternatives, so the cost might not be justified.

Fatty acid esters of carbohydrates, such as sorbitan mono-oleate have been used as emulsifiers and are also potential enzyme synthesis products.

Stabilizers and Gelling Agents

Emulsions are inherently unstable and are liable to break down, either by creaming—the phenomenon seen in milk where cream slowly accumulates in a floating layer—or by flocculation and phase separation. It has been found that addition of water-soluble polysaccharides can prevent or slow down these effects. There are broadly two mechanisms involved. The droplets are prevented from coalescence by repulsions though inevitably some contacts, due to thermal agitation, will overcome the repulsive force and lead to coalescence. Thus the emulsion will slowly break down. Added stabilizers work in two ways. Some are adsorbed to the surface and effectively raise the contact energy required for coalescence. They are in fact emulsifiers in their own right. Others can be seen as primarily raising the viscosity of the medium, so that contacts are less energetic, and sedimentation is slower. In the extreme such systems gel so that the lipid droplets are held within a fairly rigid network out of contact with each other.

Table 5.1 lists the main polysaccharides which are used in foods, together with their source and the monosaccharides they contain. In many respects the functional properties of polysaccharides are similar to those of proteins, which of course they resemble in being long-chain polymers. There are, however, some fundamental differences. The number of different monomer units is much smaller. Proteins contain all the 20 different amino acid residues, while polysaccharides rarely contain more than four different ones, and some, like dextrans, are actually homopolymers. Also, although many of them bear negative carboxyl or even sulphate groups, they can be neutral, which proteins never are. But the most significant structural difference is because they can have branched structures of various kinds. This affects their rheological and gel-forming properties. The polymers can interact, where their structures permit, by forming helical junctions or by ionic interaction, particularly involving Ca^{2+}. In such cases gels form, often at low-weight concentrations. In other cases, the viscosity of the solutions rise as the concentration increases but they do not gel. Mixtures of different polymers can show complex interactions, with non-additive effects on the viscosity.

Table 5.1 Carbohydrate polymers used in foodstuffs as stabilizers, emulsifiers and gelling agents

Name	Monomer units	Source
Pectin	D-galacturonic acid (some methylated), L-rhamnose, D-galactose, D-arabinose	Fruit
Gums		
Acacia (Arabic)	D-galactopyranose, L-arabinose, L-rhamnose, D-glucuronic acid	*Acacia* sp.
Locust bean	D-mannose	*Ceratonia siliqua*
Tara gum	D-galactopyranose	*Caesalpina spinosa*
Tragacanth	D-galacturonic, D-xylose, L-fucose, D-galactose	*Khaya* sp.
Ghatti	D-mannopyranose, D-glucopyranose, D-galactose, L-arabinose	*Anogeissus latifolia*
Guar	D-mannose, D-galactopyranose	*Cyamopsis tetragonoloba*
Algae		
Agar agar	D-galactopyranose, L-(3→6)-anhydrogalactopyranose	Red seaweed
Alginic acids	D-mannuronic, L-guluronic	Brown seaweed
Carrageenans		
Furcellaran	D-galactose-(2→6)-sulphate, D-(3→6)-anhydrogalactose	Red algae
Bacteria		
Xanthan	D-glucose, 6-acetylmannose, glucuronic acid, pyruvyl mannose	*Xanthomonas campestris*
Pullulan	maltotriose units	*Aureobasidium pullulans*
Curdlan	D-glucose	*Alcaligines faecalis*
Dextrans	D-glucose	*Leuconostoc mesenteroides*
Gellan	D-glucose, D-glucuronic, L-fucose, some acetyl groups	*Pseudomonas* sp.
Bacterial alginates	As alginic acids	Numerous mucoid bacteria

For all these reasons there has been much interest in the solution properties of edible polysaccharides, and they have been widely exploited in foods as minor ingredients which make a large contribution to the texture as well as for their emulsion stabilizing properties. There are a number of possible applications of biotechnology techniques.

PECTINS

Pectins occur in the cell walls of fruits. Commercially they are byproducts of cider manufacture, and derive from apples, or in non-cider making countries from waste citrus peel. Pectin has a branched structure, and gels readily through association of the long linear regions. Ca^{2+} assists in gel formation, though this depends on the degree of methylation, which varies with the origin of the pectin. As is well known, the structure of jams and preserves depends on pectin, and commercially it is usual to control the level, if necessary by adding isolated pectins. The rate of gelation is regulated by using pectins with higher (fast) or lower (slow) methylation. Pectin esterase (EC 3.1.1.11) is widely distributed and removes methyl groups from pectins. It is present in fruit, but interestingly is inhibited by high sucrose levels. This no doubt accounts for the stability of jams, which contain it. Pectin esterase produces methanol, as a result of hydrolysis of the methyl groups and this can give rise to some problems in citrus fruit juices. There are a number of other enzymes which break the glycosidic links in pectins, and in processing terms affect the viscosity, sometimes desirably and sometimes not. Thus they are used to clarify fruit juices, and reduce the viscosity of tomato sauce. On the other hand, they are responsible for an unwanted settling out in turbid citrus fruit juices. Commercial pectinases (EC 3.2.1.15) are made from *Aspergillus niger* cultures, and are among the top ten enzymes in production. They are used on a batch basis, and so far immobilized forms have not been used, though they have been made. Commercial preparations contain both the esterase and pectin lyase (EC 4.2.2.10), as well as the hydrolase, and have found major application in hot-pressing processes for extracting juice. In these, the fruit is crushed, the enzyme added, the mixture heated and pressed and the juice filtered in one operation. As we have seen already for other processes, the thermal instability of the enzymes is crucial since their activity must be stopped at the right moment. There is some indication that the esterase and the lyase must both be present to obtain good clarification, but an enzyme is known which can hydrolyse the methylated pectin directly. This might have some advantages, but on the whole the technology is well established and improvements are likely to involve better control of the relative proportions and amounts of the different enzymes rather than novel ones. The pectic acid lyase from a carrot rot organism, *Erwinia carotovora*, has been cloned in *E. coli* but is not yet established as a production system. The intended market for the enzyme is in instant tea manufacture, as an aid to tea leaf extraction.

GUMS AND EXUDATES

Gum Arabic is a traditional material obtained from *Acacia* species and is widely used in confectionery. It does not gel. Ghatti and tragacanth are used as substitutes, depending on the supply of gum arabic, though tragacanth has good acid stability and can thus be used as an emulsifier in salad dressings. Guar gum is particularly interesting at present. It is one of a group, which also includes locust bean gum with a linear mannan core, and galactosyl residues as side chains along its length at varying frequency. They have found extensive use as thickeners and stabilizers, and in general do not gel. However it has been found that when the galactose side chains are removed they can form gels, though if too many side chains are taken off, interactions become so strong that precipitates result. They can then be used to replace other gel-forming gums, which at present are much more expensive. Guar gum, which comes from seeds and pods of *Cyamopsis tetragonoloba*, a tree cultivated in India, is available in quantity and during germination the pod contains an enzyme, α-galactosidase (EC 3.2.1.22) capable of removing the $(1\rightarrow6)$-α-D-galactose side chains. Attempts are now underway to transfer the relevant gene to a producing organism, probably a yeast, so that it will be available for a commercial process.

SEAWEED POLYSACCHARIDES

Agar is extracted from red seaweeds, and commercial samples may be combined extracts from several species of *Gelidium* and *Gracilaria*. It is always a mixture of agaropectin, with branched chains and the linear agarose. The latter is the main gelling agent. Alkaline treatment is used in extraction and seems to remove sulphate groups, and at the same time enhance gelation ability. There is some seasonal variation in the quality of agar, probably connected to the average molecular weight. The gel is widely used in Japan as part of fruit desserts and in *tokoroten*, a soy sauce flavoured item apparently similar to the Welsh laver bread, also made from seaweed.

Carrageenins are obtained from red algae, and are widely used in small amounts because they can be gelled with Ca^{2+} and because they react with guar to give a range of gels. Ice-cream and yoghurt owe much to carrageenin. They have also been used in imitation caviar work. Lipid droplets, flavoured with fish roe and encapsulated with the polysaccharide are claimed to provide convincing results. A large number of patents were filed in the USSR on this technology: the sturgeon is said to be becoming scarce.

Similar mixtures can be used to make imitation blackcurrants by the encapsulation of juice, and reconstituted fruit pieces have been marketed based on the use of mixtures of carrageenins and xanthans to provide the texture. These were successful because fruit processing always creates small pieces and pulp which cannot be used in the main product, and has little value. Reconstitution raises the value considerably.

Carrageenins are not in plentiful supply, because although they are extracted from seaweed, the right species does not grow everywhere and is not easy to harvest. Thus prices are high, though since the amounts added are small they are much used. However the potential shortage has led to a hunt for other sources with similar properties.

ALGINATES

Alginates are extracted from brown algae such as the giant kelp, *Macrocystis pyrifera*. They are block copolymers of mannuronic and guluronic acids, and vary slightly from species to species. They sometimes become in short supply and farming of the algae as opposed to simply collecting it is likely to develop in the near future.

There is an enzyme, an epimerase which converts mannuronic to guluronic acid, and can do this to the polymer. This offers an opportunity therefore to control the ratio and distribution. It has not yet been used for this purpose.

XANTHANS AND MICROBIAL GUMS

The microorganism *Xanthomonas campestris* is the source of xanthan gums, which while they do not gel, can form gels when they are mixed with debranched guar or carob gums. They are used in much the same range of products as the traditional gums. As a fairly recent introduction, xanthans have been subjected to toxicological investigation, and are accepted for food use by the US Food and Drug Administration.

The success of xanthans has inspired a number of other attempts to use bacterial gums. A good deal of work was inspired by their non-food uses as well, since they are used in oil drilling as a flow aid. Pullulan from *Aureobasidium pullulans* is used to make non-digestible sheets for incorporation into cod-roe layered materials. Curdlan, from *Agrobacterium*, forms gels and has been used in Japanese desserts, and in low-calorie salad dressing. Like a number of these microbial polysaccharides, it is not broken down by digestive enzymes, and may even resist gut bacteria as well.

Konjac is a glucomannan from *Amorphophallus konjac*, and is used in a wide range of Japanese meals. Dextrans, including cyclic dextrans are also bacterial, originating in *Leuconostoc* by an interesting transferase reaction in which the glucose moiety of sucrose is added to a polyglucose chain, releasing fructose.

CONCLUSIONS

If we add to these, the chemically modified starches, such as carboxymethylcellulose, it is clear that there is a tremendous range of polysaccharides which separately or in combination can be used to give almost any desired texture to a food. In the next few years the relationship between structure and functional properties for these molecules will almost certainly be exploited not so much to increase the variety of textures that can be made, but to ease supply problems. The main impetus behind the development of new gums is the potential shortage and price of existing ones. Thus, one gene-transfer development which is being done now is aimed at a cheaper source rather than a novel material, though that too may happen. Another factor which should not be overlooked is that many of the novel products, although they are apparently freely used in Japan, have not yet been approved for food use elsewhere. Microbial products will always have to surmount this hurdle.

Flavours and Enhancers

Many of the added flavours used in foods are simply extracts of the natural source. Prawn and chicken flavours for example are extracts from unwanted bits and pieces—byproducts of normal processing. Others are specific substances, often relatively simple organic compounds, which are accessible by a biotechnology approach. Flavours are vital to formulated foods—perhaps the most sensitive are polysaccharide dessert-type products—and most food manufacturers maintain a laboratory that analyses natural sources for flavour components, as well as competitors' products. Formulations are closely guarded secrets. Natural flavours are the result of hundreds of components, often present in trace amounts, acting together, though a few dominant ones can usually be recognized. The general aim is to provide acceptable flavours, in a competitive situation. A new and improved flavour can lead to a significant increase in market share, while lack of it can have the opposite result, more directly than any other variable in a food product. There are some non-food uses for flavours as well. In the UK they like peppermint in their toothpaste, though elsewhere they like oil of wintergreen. Although the flavours used in foods vary widely round the world there is no real data on preferences, though there are some nearly universal likes and dislikes.

LACTONES

Lactones are important flavour components that are synthesized by microorganisms. γ-Decalactone can be made by growing *Candida* on ricinoleic acid (from castor beans) where it makes the 4-hydroxydecanoic acid. Boiling the culture medium converts this to the lactone. A number of other fungi can make lactones or their precursors, which are generally responsible for the fruity, and even coconut aromas associated with them. No production methods have yet been developed based on them.

ESTERS

Esters such as ethyl acetate and butyrate are important in fermented drinks and are made by the yeasts used. Most yeasts only produce trace amounts, but *Hansenula* and *Candida* species are known which make potentially useful levels.

The use of isolated lipases for ester synthesis has already been mentioned above. A process for geranyl and citronnelyl butyrate has been described, though it is not clear whether it is still in use. One interesting observation is that ethyl propionate or acetate are poorly formed by lipases. Lipases generally prefer long-chain fatty acids. However in the presence of small amounts of long-chain acids, the yield of short-chain ones was improved and it is possible that inter-esterification mechanisms come into play.

TERPENOIDS

Numerous microbial transformations leading to the interconversion of terpenes are known. α- or β-pinene can be converted to L-carvone which is valuable in peppermint flavours. Menthol, which is obtained as an extract of mint, crystallizes as a D,L-mixture and resolution of this with the aid of enzymes has attracted a lot of attention. Microbial esterases will preferentially break the L-methyl esters, leaving the D-esters behind. The L-form is the one required in flavourings. A lipase from *Candida lipolytica* has been used to preferentially make a lauryl ester of the L-form. The ability of enzymes to distinguish between stereoisomers is often cited as one of their main advantages over chemical reactions, but examples in food use are rare. This appears however to be one actually in use.

The Use of Plant-Cell Cultures

When plant-cell culture first became feasible about 20 years ago it was seen as a possible route to flavours and colourings for foods. It seemed obvious that the relevant cells could be cultured. One project started then became transmuted over the years into one for the clonal production of pine trees, for pinene production, thus in a very indirect way reaching one of its objectives.

For the most part, the original targets have not been reached yet. There are two principal reasons for this. One is that plant cells divide much more slowly than bacteria—20 h as opposed to 20 min—and no-one knows why this should be so. Probably the single factor which would change the commercial prospects of plant-cell culture would be to increase the division rate. A second reason is that, although it has been found possible to grow plant cells in similar culture vessels to bacterial cells, large-scale fermentation is very expensive compared with agriculture, and the competition in this case is direct. A minor reason is that plant cells do not secrete material into the medium very easily, making recovery of the product more difficult.

Similar considerations apply to algae, with the added complication that they require intense light supplies, which are not usually available from natural sunshine. They have been considered as sources of high-value colours, as well as polysaccharides, but costs are high. Despite all this, shikonin, a colouring material, and rosmarinic acid have been made in culture, and some pharmaceutical applications will keep the technique alive. A process to make capsaicin, the hot component of chili, has been considered for scale up, while saffron has also presented a useful target. One interesting technical possibility is the development of plant hybridoma cells, analogous to the monoclonal antibody production systems. A related development, though not strictly plant culture, is the growth of a yeast containing astaxanthin. This may one day find a market, since the colour is used to make sure that farmed salmon are pink. The pink colour which derives from shrimps and other Crustacea in the normal diet, and actually is astaxanthin, has to be added to farm diets. There is undoubtedly some consumer resistance to grey smoked salmon.

CARBOXYLIC ACIDS

These are by far the most important food ingredients made by microbial fermentation. The tonnages used annually are large and run into tens of thousands

(Table 5.2) though they are not all used in foods. Gluconolactone is used, but most of it is actually employed in bottle-washing plants as a general antiseptic.

The acids listed in Table 5.2 are found widely in all fruits, with the exception of tartaric (mainly found in grapes—and thus wine—and tamarinds). They do not have any flavour themselves but contribute markedly to the characteristic flavour of the fruits by providing sour and sharp or bland backgrounds. Glutamic acid is well known as a powerful flavour enhancer, particularly in chicken products.

Apart from this the major use of carboxylic acids is in regulating the pH of fruit juices, gelatin gels, canned fruits, wines and ciders. In some dried mixes, the lactones are used as latent sources of acid. In confectionery, they are used as flavour adjuncts but also to 'invert' the sucrose. They can all be made by chemical synthesis, in the D,L-form, and this is actually the version of malic acid most widely used. It is cheaper than the biosynthesized L-form, and is permitted in most countries. On the other hand, L-tartaric acid is more soluble, and preferred to the more expensive synthetic acid. Commercial lactic acid is usually a mixture with about 60% of the L-form—different bacteria vary in the presence of racemizing enzymes.

Production of Carboxylic Acids

Citric Acid. Large-scale production is based on the use of *Aspergillus niger* grown in glucose or sucrose, and is carried on in most countries in Europe and the USA. In the oldest established process the fungus is grown in surface culture in diluted molasses. It is interesting that a problem with copper, zinc, iron and manganese levels in molasses can be solved by adding ferrocyanide. This locks up the ions and makes them unavailable to the microorganism. In order to maximize the citric acid level, the activity of the enzyme aconitase, which breaks down citric to aconitic acid, must be minimized, and these ions activate it. The strains of *A. niger* used are chosen, naturally, to be high citric acid accumulators, and this is probably due to their low levels of both aconitase, and isocitrate lyase, both of which remove citric acid.

There are two kinds of fermenter in use: one where the organism is grown in shallow trays, and a more recent introduction, a conventional stirred fermenter. Both eventually produce a culture medium from which citric acid is obtained by

Table 5.2 Annual production of carboxylic and amino acids by fermentation

Acid	Tonnes	Source
Citric acid	300 000	*A. niger*
Gluconic acid	50 000	*A. niger*
L-Lactic acid	20 000	*L. delbruckii*
L-Glutamic acid	220 000	Synthetic or *Aspergillus* sp.
L-Tartaric acid	40 000	Wine byproduct

crystallization after clarification, by adding calcium hydroxide and charcoal to adsorb colours. Some manufacturers may use solvent extraction with butanol, or long-chain amines in butanol.

Yeasts can also be used, and a wide variety of monosaccharide-containing substrates have been investigated. So have the potential uses of entrapped and immobilized cells. A large-scale fermentation process such as this one will become adjusted to suit local conditions and many variations have appeared: it will also obviously stimulate a good deal of laboratory development work. However, the existing methods have been in use for some years and there is no sign of any significant contribution from genetic manipulation techniques for the time being.

Glutamic Acid. Strictly speaking, this is an amino acid, but it is convenient to consider it here. It is the major flavour enhancer, and is made in huge amounts, particularly in Japan. It can be obtained from hydrolysates of seed storage proteins such as soy protein, which are rich in it. About 20% of the residues are glutamate or glutamine. In practice, however, it has proved cheaper to make it by fermentation of molasses supplemented with a suitable nitrogen source, with *Corynebacterium glutamicum*, which secretes at least 40 g/litre into the medium. It is usually isolated from the medium by an ion exchanger, followed by crystallization as the sodium salt.

Lactic Acid. Lactic acid is made by the anaerobic fermentation of sucrose or glucose, in the form of molasses with *Lactobacillus delbruckii*, though whey, or rather the lactose in it is some times used with *L. bulgaricus*. The culture is first clarified by adding calcium hydroxide and filtering. This yields a solution of calcium lactate together with monosaccharides and other soluble components. Recovery of lactic acid is not easy, because concentrated solutions of lactic acid tend to condense through the hydroxyl and carboxyl groups, and do not crystallize. They can also be difficult to separate from monosaccharides. Solvent extraction is probably the best method for general use, followed by ion exchange for final purification. However, other methods, such as esterification followed by distillation and hydrolysis have also been proposed. This is a case where a more expensive culture medium may more than save its costs in the final purification stages.

Since L-lactic acid is the desired form, it is important to choose organisms that reduce pyruvic acid to lactate in a stereospecific manner. One is known that has two lactate dehydrogenases with either L- or D-specificity. This is a remarkable observation, and it would be interesting to know more about the relationships between these two enzymes. It could give an important insight into how to obtain such specificity more generally. There is clearly scope for genetic manipulation to combine other desirable qualities with the most appropriate dehydrogenase.

Lactic acid is the main preservative in a number of fermented foods, of which sauerkraut and some sausages are the best known. These usually employ traditional mixtures of *Lactobacillus* strains. The process is monitored, or should be, by measuring the pH.

Gluconic Acid and Gluconolactone. Gluconic acid is produced by *A. niger* grown on glucose in large stainless-steel fermenters. The pH is maintained at 6.5 with sodium hydroxide, and eventually a mixture of the sodium salt and the free acid are obtained by evaporation.

Gluconolactone is the main form in which gluconic acid is actually used and is made from the free acid. The sodium salt is most easily converted by using ion exchangers. Although it could then be turned into the lactone by using either a glucose oxidase or dehydrogenase, in practice the lactone is obtained from spontaneous conversion of the acid in concentrated solution, by crystallization in defined conditions.

Conclusions

There is now every reason to expect that lipases, and possibly other esterases will become as well established in lipid processing as are amylases in starch derivative manufacture. Their role in flavour and emulsifier processes is also likely to be significant, though here the limited size of the markets for any one ingredient may limit the rate of progress. As was pointed out in Chapter 1, the cost of gene-transfer techniques is difficult to justify unless the components of interest have a sizeable cash flow attached to them. In the example of lipases, the availability of purified ones needed for non-food use is a fortunate circumstance which will not always happen. It is clear that there are many opportunities for modification of polysaccharides of food use, but the main impetus behind current research is to obtain more reliable and cheaper supplies. The market size for many of these components is not enough to justify full-scale genetic engineering, and progress will probably have to depend on chance discovery of suitable enzymes.

Chapter 6

Food, Populations and Quality Control

Introduction

This chapter is primarily concerned with hazards associated with foodstuffs and the part which biotechnology might play in reducing them. There are three aspects to be considered.

The first aspect is the rather obvious one of ensuring that the label is correct and that the food is free of dangerous fungi and bacteria—the food-poisoning organisms. It used to be said that the food industry is really a matter of packaging and preservation, and there is still much truth in this. Both the requirements of legislation and a proper service to consumers require clear accurate labelling. This in its turn calls for adequate analytical methods, for without these rules cannot be enforced. At one time certain countries used to have extremely rigorous 'food purity' laws, but since they had no adequate laboratory facilities within their boundaries, and the actual practice was very different from the rules, this was little better than a propaganda exercise. On the other hand government laboratories have been known to make mistakes in their analyses. It is not surprising that most large food manufacturers maintain analytical laboratories to monitor both their own products and incoming raw materials. Biotechnology has made significant contributions to analytical methods, in the use of immunological techniques, and in detection of bacteria.

The second aspect is concerned with the very extensive range of toxic materials that occur in foodstuffs and potential foodstuffs. When one surveys the numerous toxic legumes it is rather surprising that any prove to be edible at all. Biotechnology has contributions to make to detoxification methods.

The third aspect is slightly less obvious but is concerned with the heterogeneity of human populations. It tends to be assumed by those who set standards that they are dealing with a homogeneous population but we now know that they are not. Digestive enzymes present in one population may be absent in another. As people travel more widely, and the foods and habits of one continent become quite rapidly transferred to another, unanticipated risks may be present. Populations and their major food plants have probably evolved together and changing the balance may lead to unexpected effects. 'Montezuma's revenge', a particularly virulent form of Mexican food poisoning is well known and immediately obvious. There are many less evident perils of changing one's food environment. In the past, changes in agricultural practice and food supply have been seen as precursors of population increase. The situation is now reversed, at least temporarily, and mankind is engaged on a huge experiment, under the pressure of population increase, in novel foodstuffs.

Testing and Hazards

Over the last 200 years, mainly as a result of attempts to preserve foods, a number of 'additives' have come into use. They vary greatly, from common salt to elaborate dyes. For example, canning processes often change the colour and it has been customary to use added dyes to restore the colour to something approximating that of the fresh food. Although many of these additives have been in use for long periods in recent years they have been challenged on the grounds that the potential risk to consumers, even if it is very small, is not justified by any significant gain from their use. This is probably true for some colouring materials, and in countries where they are no longer permitted there has been no noticeable drop in market size attributable to a change in colour. The matter is rather different when preservatives are questioned since there is a very real connection between their use and food-poisoning incidence. The balance between well-established advantage in the use of a substance like nitrate in meat curing, as against an unquantifiable but theoretically possible risk from this substance, will always be controversial. No-one can ever prove the complete absence of risk, and anyone can demand it.

It is now certainly the case that any new substance or process that may alter the composition of a foodstuff will be subjected to close examination in most countries. Both the USA and the EEC lay down very detailed requirements. This is particularly the case when the products of recombinant organisms are in question. The major requirement in all such regulations is the containment of the transformed organisms. As more experience has been gained the list of organisms subject to the highest degree of containment has become shorter, but it must always be remembered that a proposal to produce an enzyme in a microorganism will have to satisfy the local regulatory authority, which can have unexpected views. In some parts of Europe, all work on recombinant organisms was prohibited while a short distance away, across a frontier a different view prevailed. Attempts to cultivate transformed plants, except in high containment greenhouses, have also met strong resistance. The regulatory response to proposals to cultivate uncontained transformed plants remains a major uncertainty. One group has gone so far as to make a fluorescent strain of *Pseudomonas* so that it could be easily traced in field trials.

Handling enzymes in the industrial context has also led to difficulties. It was quickly found that large quantities of dry enzyme powders pose considerable hazards to the workforce, and even in laboratories regular monitoring of people handling them is desirable. Unless precautions are taken a high proportion of people handling microbial enzyme powders will become sensitized by the inhalation route. In factories, enzymes are now used in liquid suspension, to avoid dust whenever possible.

Apart from these problems of manufacturing operations, enzymes have also been scrutinized as potential 'additives', since they count as either components or process aids. The main problem facing regulators is that, 'enzymes', as used in industry, are in fact very complex mixtures derived mostly from microorganisms. In the UK, three lists have been produced. One contains such organisms as *B. subtilis*, *A. niger*, *Rhizopus* and *Mucor*, and *Saccharomyces* and *Kluyveromyces*, all of which from long experience are belived to be in no need of further testing. A second list, known to occur in foods without harmful effects are thought to need some testing, while a third one contains organisms that are known to be undesirable. Any product originating from these will be subjected to very extensive testing. In the present climate of opinion organisms about which little is known, and transformed organisms, will probably have to be treated as in the third, riskiest category.

Biotechnology has made little contribution to toxicity testing methods. Molecular biology is increasingly being involved. Mutagenicity is measured in a standard test in *Salmonella* actually by determining the rate of back-mutation induced by the substance in question, in a mutant form. Work is also done on the direct interaction of materials with DNA, and there are possibilities of using transformed organisms in tests for specific effects on enzymes. Almost all toxicological testing is however done on whole organisms, usually mammals.

ANALYTICAL METHODS USING ANTIBODIES

The most familiar situation where antibodies are used is where adulteration is suspected. In recent years, for example, there has been much concern that vegetable proteins, such as soy, might be illicitly incorporated into meat products. Since many meat products do contain some soy protein, and declare so on the label it is not sufficient merely to detect soy protein. A method capable of estimating its amount is also needed. One meat-producing country actually banned its use and importation on the grounds that it presented an irresistible temptation to its citizens, which could not be detected by any existing method. This gave an added impetus to the search for one.

Another example is where the meat content of a product, perhaps canned meat stew, is not of the species specified. In one celebrated case kangaroo meat was sold as beef: kangaroo is available in some quantity in Australia, and is used in petfoods, but should certainly not be sold under a different label. There are many other examples, from the legendary cat sold as rabbit, to the undeclared use of chicken meat, now that it is relatively cheap.

In Chapter 1 the great resolving power of electrophoresis was mentioned. An electrophoresis pattern is virtually a map of the proteins, and thus indirectly of the active DNA, and can certainly distinguish between species and individuals. It is however completely dependent on getting the sample into solution, and it is rather laborious to obtain quantitative precision. Although many attempts have been made, not all unsuccessful, the use of antibodies has proved more practical than electrophoresis. The direct use of DNA probes, although clearly able to identify any species, is not likely to come into use for several years. It has already been used in experimental work to detect bacteria, and probes for *Salmonella* and *E. coli* have been prepared. Detailed analysis of the plasmid patterns has also been used to identify strains of *Salmonella*, and was able to show that in one case of *Salmonella* infection, the source was a sparrow living in the factory roof. It is not at all clear which method will prove to be the most convenient in practice, and for the moment antibodies are the most widely used.

All vertebrates produce antibodies, and when a protein is injected into a suitable animal, and mice, rabbits, sheep, goats and horses are used, the latter mainly in commercial antibody preparation, it produces immunoglobulins which characteristically form complexes with the antigen. Each of them has two binding sites, so that infinite chains can form, and precipitates. The formation of precipitates in agar gels, as antibodies and antigens diffuse towards each other, meet and react is widely used in analysis.

Such an antiserum is described as polyclonal, because it contains a mixture of immunoglobulins, binding to a variety of different sites on the antigen. Because of this variety they usually show the phenomenon of cross-reaction. Thus, if an antibody raised against glycinin, the major storage protein of the soy bean, is reacted with arachin, the main storage protein of the groundnut, some of the antibodies, but not all, will react with it. The two proteins are actually fairly closely related, and the same cross-reaction, or reaction of partial identity can be seen throughout the legumes. There would be no problem in using such an antibody to detect legume protein in meat—there is no cross-reaction with meat proteins. This is also a situation where a polyclonal antibody may have advantages. Because it reacts with numerous sites on the protein, it will still have a good chance of reacting even when the protein has been autoclaved in the presence of many other components.

Quantitation is not simple, but automated techniques have been developed, like the ELISA system. The basis of this is illustrated in Fig. 6.1 and can be used for all antibody tests. It is much more sensitive than the precipitation techniques.

As can be seen, polyclonal antibodies are not specific, but their specificity can be improved by the technique of adsorption. In this, the antigenic proteins, often attached to a column, are used to pull out the antibodies, which are of course present in a complete serum containing all the other serum proteins, and any other antibodies already present in the animal. They can then be eluted and if necessary further adsorbed to remove unwanted antibodies. For example, an anti-soy protein antibody could be adsorbed with groundnut protein, so that the remaining

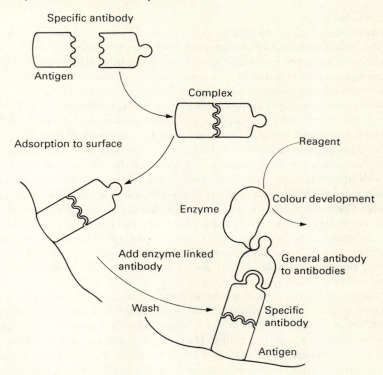

Fig. 6.1 ELISA, enzyme-linked immunosorbent assay. The products of reaction between the antigen and a specific antibody are detected and quantified by reaction with a second antibody against immunoglobulins in general. An enzyme covalently linked to this then generates a coloured product which is estimated by automated spectrophotometry. Alkaline phosphatase, galactosidase or peroxidases are used. It is also possible to label the second antibody with ^{131}I making it a radioimmunoassay, but non-radioactive methods are preferred.

antibodies would not then be capable of recognizing groundnut. Meat proteins present difficulties. Actomyosin is so highly conserved (that is, it is so similar in all species) that it is difficult to obtain an antigenic response—animals do not normally make antibodies against their own proteins. Even when antibodies can be obtained, cross-reaction is extensive and antibodies against actomyosin cannot be used in analysis. The sarcoplasmic proteins are most often used, and also serum proteins, since these are present in meat. By means of adsorption, the use of antibodies against antibodies and cross-reaction, it is possible to devise all sorts of complex schemes most of which can be made to work under laboratory conditions. Routine analysis is a different matter. The success of the ELISA technique depends on the unpredicted ability of the antibodies to stick to plastic surfaces.

Monoclonal Antibodies

These have attracted much attention because of their potential uses in pharmaceuticals, but as a result have become available for some food analyses.

Myelomatosis is a relatively uncommon condition in which tumours of plasma cells result in over-production of immunoglobulins. It was realized that since the tumour originated in a single cell, only a single antibody was being made. Some early work on sequencing of immunoglobulins both established this, and depended on this chance supply of a pure antibody. The antibodies were of course different between different individuals. The same condition has been found in mice and rats. The basic operations for the production of monoclonal antibodies are given in Fig. 6.2. Essentially a mouse is immunized, and then the antibody-producing cells fused with myeloma cells, which are then allowed to proliferate, and selected so that each culture originates in a single fused cell, and thus produces only a single antibody. These are screened with the original antigen, to select those that are making antibodies against that protein. Many others will also be present from the mouse's normal antibody production. In the original methods it was necessary to culture the cells in a mouse peritoneum, which severely limited the quantity available, but this has now been replaced by *in vitro* cell-culture methods and kilogram quantities are now made.

Although it is relatively easy to obtain reacting antibodies, the strength of the interaction is often very weak, and several hundred monoclonal lines may have to be tried before a suitable one is found. Suitability is assessed on the basis of specificity and strength of interaction. If a suitable one is found it then provides a powerful tool and can be used in the ELISA methods. It should be emphasized however that there have been a number of (unpublished) attempts where suitable monoclonals were not found. An advantage of monoclonals, now that tissue-culture methods have been developed, is that they are commercially available in unlimited amounts, and are consistent, which was never the case with polyclonals. This is such an advantage when it comes to routine operations that monoclonals will probably be used, even though they have no other advantage over polyclonals. A moment's thought will show that monoclonals can never be precipitating antibodies unless at least two of them are mixed. Gene manipulation methods can be applied as well to immunoglobulins as any other proteins, and as the basis of specificity in the antigen–antibody reaction is understood will almost certainly lead to a supply of novel antibodies, which may find application in this field.

AFLATOXIN ANALYSIS

This is a good illustration of the use of antibody methods for screening for dangerous contaminants. Mycotoxins are now extensively tested for, mainly because of an outbreak of aflatoxin poisoning in 1960. In that year, large numbers of turkeys died suddenly in the UK, and after eliminating the first idea that it was due to a novel virus disease, the cause was traced to a consignment of groundnut meal imported from Brazil, a novel source for UK feed formulators. This was found to contain a series of closely related substances—the aflatoxins—which are

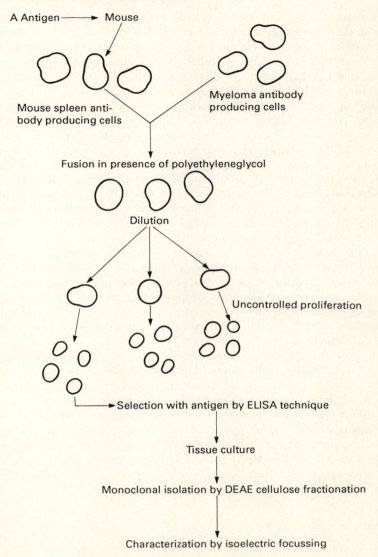

Fig. 6.2 Schematic monoclonal antibody production.

secreted by *Aspergillus flavus* (and also *A. parasiticus*). It was soon found that these substances could be found in a variety of seeds and grains, which had been allowed to go mouldy. Inadequate storage in tropical conditions was identified as the principal cause of contamination. Apart from acute toxicity—and poultry are particularly sensitive—there was evidence that long-term exposure causes primary hepatoma. This condition, in man, is very rare in temperate climates, but is much more common in the tropics, and its possible association with aflatoxins present in the food was soon pointed out. There is certainly a strong correlation between the incidence of primary hepatoma and the level of aflatoxin in the food in large parts of Africa. A number of outbreaks of aflatoxin poisoning were reported in Africa and Asia, and in some studies it was found to be strongly associated with the symptoms that had previously been ascribed to kwashiorkor. The matter is not resolved, but it appears that kwashiorkor may be more than a simple protein-deficiency syndrome. It may of course be simply a reflection of the likelihood that poor storage of food will coincide with areas of food shortage. It has also been suggested that malnutrition may weaken defences against low levels of aflatoxins. In Thailand, mycotoxicosis has been identified as a common cause of death in children. The position in Europe and the USA is controversial, since aflatoxin has been suspected of causing the rare Reye's syndrome, but this is, at best, not proven. It is however quite indisputable that aflatoxins are highly undesirable substances to be present in food, and international guidelines have been set. The limits vary and are set at $20\,\mu g/kg$ for peanuts (which are the same as groundnuts or monkeynuts) in the USA. Most manufacturers monitor oil-seed meals routinely, but one effect of the original discovery of aflatoxins was the abandonment of projects to use groundnut protein in meat products for human consumption. Trials have shown that, even in the tropics, proper storage and handling generally can keep the aspergillus infection down to acceptable levels. Soy meal used for protein extraction for human consumption mostly originates in the USA where storage facilities are excellent.

It is obvious that a great many analyses for aflatoxins are now done, and one of them uses an antibody. Antibodies to small molecules like aflatoxins are raised by attaching them to proteins in the so-called 'hapten technique'. The particular variation used for aflatoxins consists of an immobilized antibody, which is exposed to the aflatoxin-containing sample. This interacts with the binding sites leaving some vacant at an appropriate dilution. Then, an aflatoxin attached to horseradish peroxidase is added and occupies the remaining sites. Excess is washed off, and then colour developed using 4-chloronaphthol, which forms a blue insoluble precipitate with the peroxidase. The more colour, the less aflatoxin in the original sample.

DETECTION OF BACTERIA

There is a continuing need for rapid and sensitive tests for the identification and enumeration of bacteria in food samples. In the UK, the number of outbreaks of food poisoning of bacterial origin has been rising steadily for some years. Present methods are effective but time consuming. They depend essentially on the culture of bacteria in defined media, as a means to identification, which by its very nature takes several days to perform. In the extreme case, a consignment of food might be despatched and consumed before the manufacturer's tests could be completed.

Phages are viruses which invade bacteria: they show species and even strain specificity. They can actually be used to transfer genetic material. Although plasmids, as described in Chapter 1, are more frequently used, mainly because they are readily available in a suitably modified state, DNA can be incorporated into the phage DNA by standard methods. Disabled phage, so that it can invade but not proliferate, is an excellent means of introducing DNA, which becomes incorporated into the genome, and expressed.

In a very ingenious application of this, a rapid method for detecting bacteria has been proposed. A piece of DNA, the operon for the *lux* gene, is put into the phage. The *lux* gene is responsible for the production of the α- and β-subunits of the enzyme luciferase(EC 1.14.99.21). It also contains DNA for the three subunits of a fatty acid reductase, which converts fatty acids to their aldehydes, but requires NADH. The enzyme luciferase oxidizes the aldehyde back to the fatty acid, and in so doing emits light. The *lux* gene in question was obtained from a luminescent marine bacterium, *Vibrio fischeri* by digestion of the DNA to make a genomic library followed by cloning by means of plasmids, into *E. coli*. Selection is then simply on the basis of whether the colonies had acquired luminescence or not. It was then inserted into phages, and the phage cultures selected on the same basis.

By using a phage strain for, e.g. *Salmonella*, it should be possible to detect only *Salmonella* in the presence of other bacteria, and only live *Salmonella* at that, since NADH must be supplied, It has been shown that the amount of light emitted is directly proportional to the numbers of viable bacteria. Figure 6.3 shows some results for *E. coli* in milk.

It will take some time for this technique to be established on a routine basis. The technology of light detection is advanced and available. Even a few photons can be

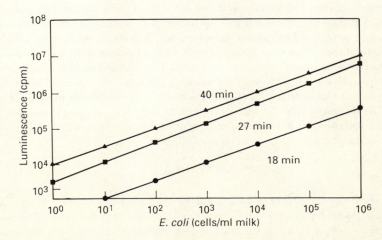

Fig. 6.3 The luminescence of *E. coli* in milk after infection with phage L28, containing the *lux* gene. The three lines are the luminescence after incubation for the indicated times. Luminescence can be used to estimate the numbers of *E. coli* present. (From S. Ulitzer and J. Kuhn (1987), *Bioluminescence and Chemiluminescence. New Perspectives.* Ed. J. Scholmerich, R. Andresen, A. Kapp, M. Ernst and W.G. Woods. Wiley, Chichester.)

detected without much difficulty. Stable phage cultures in the variety required may not be easy to obtain, and cost will also be a consideration. However, in situations where speed is important it seems very likely that this method will come into use. It should be noted that the use of phages is a neat way of getting species specificity, but that plasmids can also be used to insert the *lux* genetic material. In a rather different, non-food application, it may be possible to do quick screening for antibiotic resistance after transforming the bacteria with a *lux* gene, since dead ones have no light emission. At least two other competing methods for the detection of *Salmonella* are near production. One relies on ELISA methods and is antibody based. In one variant, an antibody has been raised against one of the proteases secreted by *Salmonella*. In addition, in methods based on direct DNA hybridization, probes have been developed to the stage of commercial application. Both of these methods claim advantages of speed, but would both detect dead bacteria, or their relics. A similar DNA probe for *Listeria monocytogenes*, another food-poisoning organism has also been made.

These competing techniques may well find optimal application in different products.

Toxic Components of Plants

We are not here concerned with acute toxicity. There are numerous plants containing alkaloids like atropine, or strychnine, which are acute neurotoxins and such plants are never consumed nutritionally. Even so, they may reach one by unexpected routes. The green quail can eat hemlock with impunity, but anyone who in turn then eats the quail will get hemlock poisoning. There are many food plants, and potential food plants, however, which contain substances which may be harmful if consumed over a long period. This is because they are present in low amounts, or because the effects are marginal. One of the most striking features of some foodstuffs is that, while they are known to be risky, they are still consumed by people who are on the margins of starvation. Presumably, the immediate threat of malnutrition outweighs the long-term risk of damage. To put it another way, consumption of foods that may cause trouble in the late 60s is not very worrying to people when the average expectation of life is 40 years.

GLYCOSIDES

One of the best examples of this effect are the cyanogenic glycosides. Many people, if asked to name a poison, would reply 'cyanide', but very few would be aware of the extent to which it occurs in common food plants, and how vital some enzymes are to their safety.

Cassava and Zamia
Cassava, also called 'manioc' (*Manihot esculenta*) is an important source of starch in the tropics, and contains a cyanogenic glycoside called 'linamarin'. The plant originated in the northern part of South America or Mexico, and was in use before

Food, Populations and Quality Control

the Spanish invasion. Because of the way it was grown, it was simply lifted from the ground as needed and not stored. In use it is first peeled, and then the tubers ground up and allowed to stand for a day before pressing to remove juice. The juice is very toxic and was used for ritual suicide, but after boiling is used as a sauce. The pressed pulp was used widely, and still is as a substitute for wheat in bread-making, and is almost pure starch. An industrial-scale wet-milling process has been developed for the starch, sold as tapioca. The leaves are also used, but only after extensive pulping and washing.

The structure of linamarin is shown in Fig. 6.4 and the way it breaks down on hydrolysis. In Nigeria it is fermented to make *gari* with *Corynebacterium manihot*, which makes a specific linamarinase, necessary for detoxification. This is an inherently hazardous sort of process because it does not 'fail safe'. If, for example, the bacteria fail to flourish, or the wrong strain gains entry, the residual glucoside levels may be too high. For this reason the possibility of adding controlled amounts of extraneous enzyme has been explored. The difficulty is that at present the only source is likely to be another microorganism of unknown properties. There may

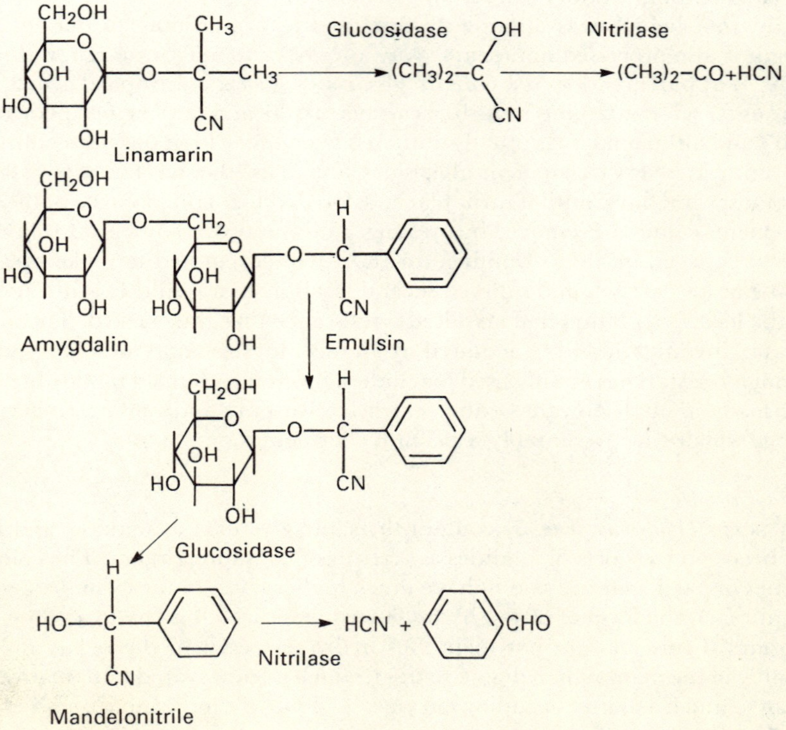

Fig. 6.4 Cyanogenic glycosides of cassava, *Zamia* and *Phaseolus* (linamarin) and of almonds, peaches and apricots (amygdalin). An indication of degradation pathways is shown.

well be a case for the provision of this, or an equivalent enzyme from a transformed source. The danger is actually lack of hydrolysis although it merely liberates cyanide. The cooking process drives it off. The risk is that unhydrolysed glycoside will enter the gut, there to be broken down by gut bacteria, with the liberation of cyanide in the wrong place. Cassava itself apparently does not possess such an enzyme, but the liberated cyanide must still be eliminated. The level of linamarin varies between varieties and it might be possible to provide low strains. The function of these glycosides is unknown, but they may be involved in protecting the plant against insect pests, so complete removal is not desirable. This was found in the case of lupins. These were grown on a large scale in Western Australia, with a view to animal food, and even for human consumption, of the seeds. They contain a number of alkaloids, which were not wanted, but removal by selective breeding led to unacceptable insect damage.

How good are the present processing methods? Not perfect, because in cassava-eating areas of Africa, a condition of the optic nerves, and a form of goitre have both been attributed to residual linamarin in the diet. That cyanide is the culprit is suggested by the presence of elevated levels of thiocyanate in the blood, which is the detoxification product of cyanide in man.

Arrowroot, which was apparently first used as an antidote to poison-arrow wounds, is also processed in the same way as cassava to make a starch, but not so widely. It apparently does not contain glycosides. *Zamia* is a tropical cycad, with tuberous starch-containing roots like cassava. Its local name means 'poor man's bread', and although it was usually gathered from the wild, it has been cultivated in Jamaica. It is very rich in toxic glycosides, and in use the starchy pulp is allowed to rot for several days, until it turns black, before baking. Unlike cassava, the toxic components cannot be removed by pressing. This was proved by a group of French soldiers, besieged in Santo Domingo in 1802, who attempted to make bread by pressing as for cassava, and suffered several fatalities as a result. It is just possible that the local population had acquired a resistance that they lacked. There have been no investigations on acquired resistance to this kind of plant poison. Although the starch is mainly used for clothes, it is still consumed in the absence of other foods, though nowadays after extensive washing and boiling rather than rotting, which was presumably a primitive fermentation.

Other Glycosides

Lima beans (*Phaseolus lunatus*) contain the same glycoside as manioc, and there have been some reports of fatalities as a result of consuming them. The coloured varieties in particular are said to have much higher levels of the cyanide, and it is easy to reach the level of 50 mg hydrogen cyanide, which is the toxic dose. The glycosides themselves are not toxic, but hydrolases exist in the bean, and also probably in the mammalian digestive tract, which release cyanide. In this instance it is not so much a matter of adding enzymes as giving the indigenous ones a chance to work, by soaking in water.

The kernels of almonds and apricots and peaches contain amygdalin (see Fig. 6.4) and of course the characteristic flavour of almonds is that of cyanide. The almonds used in making marzipan are low glycoside varieties, and are also extracted with water. There have been reports of fatalities from chance encounter with varieties of exceptionally high content. The release of cyanide needs three enzymes and there are some indications that they are highly specific, in almonds. The benzaldehyde released can also cause problems, when it is oxidized to benzoic acid. Much more important food crops like sorghum and linseed (*Linum usitatissimum*) also contain cyanogenic glycosides in the seeds. Among European foods, the common pea (*Pisum sativum*) and the black-eyed pea (*Vigna sinensis*) have small amounts—of the order of 2 mg/100 g in contrast with cassava at 110 mg/100 g—and it would be necessary to eat very large amounts to approach dangerous levels.

Small amounts of cyanide in the diet are converted to thiocyanate, which blocks the uptake of iodine to the thyroid and is goitrogenic. As has already been mentioned, the presence of goitre in the population is an indication that too much cyanide may be present in the diet. Bearing in mind that there are other causes of goitre, the incidence in most countries is evidence that cyanogenic glycosides are within tolerable limits.

Potatoes. Potatoes contain the powerful cholinesterase inhibitors solanine and chalconine. In the main potato-eating countries (the UK, USA and Germany) varieties are now screened for solanine content, and some have been withdrawn, because the content, at 200 mg/100 g, was judged too high. The alkaloid occurs mainly in the green parts of the plant and it is said that shortly after its introduction to Europe there were a number of deaths from eating the shoots. Certainly housewives nowadays excise the green shoots and eyes of potatoes before cooking them, perhaps without knowing quite why they do it.

Brassicas. Rape seed (*Brassica napus*) and the closely related mustard, have glycosides called 'glucosinolates', and also contain a hydrolase, myrosinase, which breaks them up. At least eight different glucosinolates have been described in different strains (Fig. 6.5) of which gluconapin and progoitrin are the commonest. Table 6.1 gives a useful impression of the variability of these glycosides. There has been much interest in rape seed recently because it is now more extensively grown for its lipid content. The glucosinolates however hinder attempts to use the meal for animal feeds. Methods for detoxification are being developed, and at least one of them involves the addition of endogenous myrosinase, as an aid to extraction with dilute ethanol. This might lead to a demand for the enzyme one day. Varieties with low contents have been found as well as those low in erucic acid—the so-called double low—which is expected to be widely cultivated soon. The goitrogenic effect is probably due to the isothiocyanate ions released on hydrolysis. Despite attempts over many years, rape-seed protein is unlikely to be diverted to human food, and its only use will be as mustard. Glucosinolates are widespread in *Brassica*, and are found in for example Chinese cabbages, such as *pak-choi* and *pe-tsai* (*Brassica chinensis*), but the levels are comparatively low in the UK-grown lines. Broadly, strength of flavour correlates with glucosinolate level, and it is highest in Brussel sprouts and simple cabbage. There is obviously no major hazard presented by these components at these levels, and since the flavour depends on them, removal is not wanted. The considerable variety of them suggests a complicated

Progoitrin

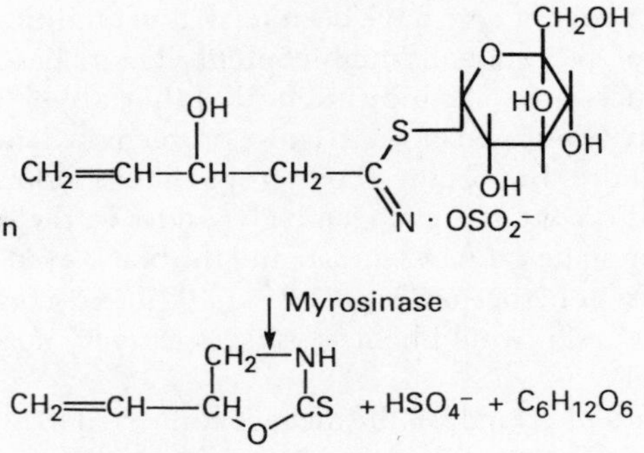

5-vinyloxazolidine-2-thione (goitrin)

Gluconapin $CH_2{=}CH \cdot CH2 \cdot CH_2—$

Glucobrassicanapin $CH_2{=}CH \cdot CH_2 \cdot CH_2 \cdot CH_2—$

Gluconapoleiferin $CH_2{=}CH \cdot CH_2 \cdot CH(OH) \cdot CH_2—$

Sinigrin (mustard) $CH_2{=}CH \cdot CH_2—$

Gluconasturtiin

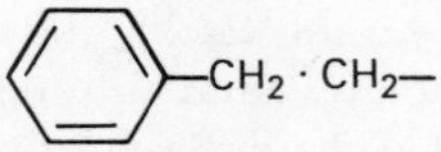

Glucobrassicin

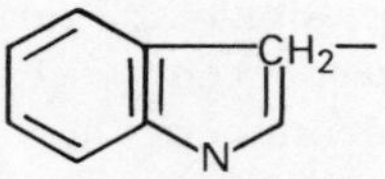

Fig. 6.5 Some glucosinolates found in *Brassica* species.

enzymology, and one day the resources of biotechnology might be deployed to control exactly which glucosinolate is present, since some of them may be more desirable than others.

Safflower contains two phenolic glucosides which can be broken down by added glucosidases, making the meal more suitable as an animal feed, and potentially available for human consumption. Similar glucosides in jojoba meal (*Simmondsia california*) can be made apparently harmless by a *Lactobacillus acidophilus* fermentation, but the mechanism is not known. The intention is animal feeding. Similar observations have been made on cotton-seed meal where gossypol toxicity is reduced by an *Aspergillus* fermentation. This may, however, be related to changes

Table 6.1 Range in concentration of individual glucosinolates in rapeseed lines

Glucosinolate		Species	Range in glucosinolate content	
			Content (μmol g^{-1} air-dried oil-free meal)	% of total
Gluconapin	I	B campestris	13–157	29–94
		B napus	3–48	12–47
Progoitrin	II	B campestris	1–85	1–59
		B napus	3–91	13–68
Glucobrassicanapin	III	B campestris	1–29	1–31
		B napus	1–30	3–30
Gluconapoleiferin	IV	B campestris	0–5	0–6
		B napus	0–6	2–5
Glucobrassicin	V	B campestris	0–1	0–1
		B napus	1–2	1–4
4-Hydroxyglucobrassicin	VI	B campestris	2–5	1–7
		B napus	3–8	3–27
Sinigrin	VII	B campestris	0–4	0–3
		B napus	0–3	0–3
Gluconasturtiin	VIII	B campestris	1–3	1–6
		B napus	0–5	0–7

From J.P. Sang & P. Salisbury (1988). J. Sci. Food Agric. **45**. 255–261.

in its physical state such as whether it is protein bound or not, rather than direct enzymatic alteration of its structure.

Bracken. Bracken (*Pteridium aquilinum*) is a very widespread nuisance in temperate climates. It is still spreading rapidly. In the UK alone it covers 700 000 hectares. It has been known for a long time that animals eating it show what appear to be thiamine deficiency disorders. It is now believed that this, together with carcinogenesis observed in experimental animals fed bracken is due to the presence of a variety of pterosins and pterosides, one of which, ptaquiloside, seems to be the precursor of the main carcinogenic activities (Fig. 6.6). There is also a thiaminase enzyme, and some anti-thiamine glycosides present. Shikimic acid, which is known to be a carcinogen in mice, has been described, and the problem with bracken is not so much to find potential carcinogens, but to decide from amongst a large number of glycosides which of them is the most dangerous.

It is all the more remarkable in view of this to find that bracken is eaten. There is some evidence that it was eaten in Europe in antiquity, but annually at least 13 000 tonnes of the rhizome is consumed in Japan, and the ferns are eaten as a salad vegetable in Canda. Processing by pickling, boiling, or salting does much to reduce glycoside levels, and the processed starch from the rhizome is usually washed clear. The fronds are probably used in many parts of the world by small isolated groups. Another problem is the way that the glycosides can appear in the milk of animals

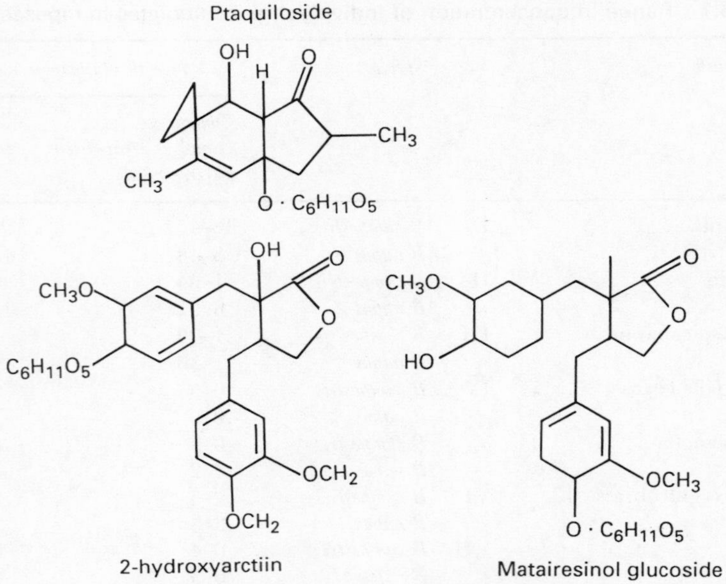

Fig. 6.6 Glucosides found in safflower and bracken.

feeding on bracken. This is particularly likely for goats, because of the marginal land they are kept on, though a study in Wales to test the idea that water supplies might be carriers failed to show any correlation between bracken infestation near reservoirs and deaths from gastric carcinoma. No problem has ever been associated with goat's milk either. While bracken is consumed on the basis of a wild plant, to be gathered freely, little can be done, but if it should be brought into cultivation then it will be possible to regulate the amount and type of glycosides it contains. There is little prospect of this happening in the foreseeable future. A systematic controlled detoxification process is more likely to be developed first.

TOXIC AMINO ACIDS

Apart from the 20 amino acids found in proteins, and a few others such as ornithine and citrulline, important in metabolic pathways, plants contain >200 others, many of which are toxic to animals. Most of them have been detected as a result of the effects they have on animals eating them, and they tend to be absent from human diets. There are however some exceptions.

Lathyrism is caused by eating the seeds of pulses, from species of *Lathyrus* (*L. odonatus*, the sweet pea) and *Vicia*. They are widely consumed in India, and unfortunately it is not easy to distinguish between relatively harmless and dangerous varieties. That some distinction is possible is suggested by the observation that lathyrism is increased during periods of food shortage. Suspicious seeds may be consumed in the absence of anything else. Osteolathyrism, which

Food, Populations and Quality Control

involves bone deformation, is caused by β-aminopropionitrile, while neurolathyrism is caused by a range including γ-diaminobutyric acid, with known neurological activity (Fig. 6.7).

The fruit of the akee (*Blighia sapida*) is consumed in Jamaica, where they are indebted to Captain Bligh of the Bounty for its presence. Now on sale in London, there have been cases of poisoning due to inexperience in its proper preparation. It contains cyclopropyl amino acids, which cause a dramatic lowering of the blood sugar. Jamaica is not entirely free from its effects, where it is believed to cause the Jamaican vomiting syndrome.

A similar amino acid occurs in California buckeye, which according to Francis

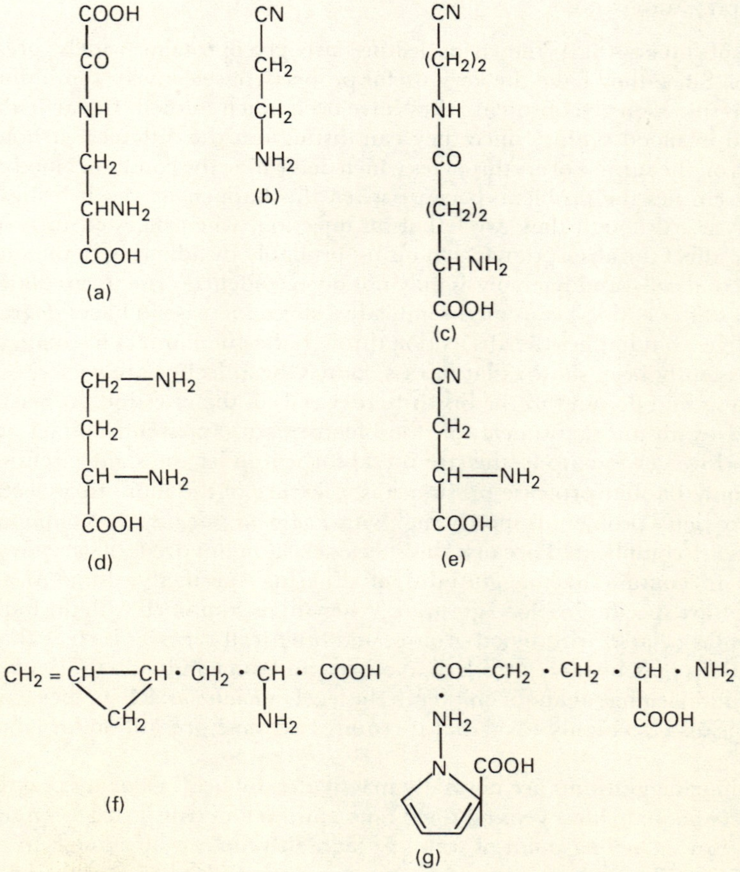

Fig. 6.7 Toxic amino acids found in plants: (a) β-N-oxalyl-L-α,β-diaminopropionic acid; (b) β-aminopropionitrile; (c) β-N-γ-glutamyl-aminopropionitrile; (d) α,γ-diaminobutyric acid; (e) β-cyanoalanine all from *Lathyrus* species; (f) hypoglycin A from akee; (g) linatine from flax.

Drake used to be eaten, but has now apparently gone out of use. Plantains contain serotonin and where heavily consumed in Africa may cause problems.

Flax seed contains a pyridoxine inhibitor, linatine, which is a glutamyl derivative of aminoproline, and must be extracted with water before the seed is fed to chicks.

It is difficult to see quite what contribution biotechnology can make to these problems. In some cases adequate information already exists about safe strains, and the problem is more one of applying existing knowledge than anything new. It is likely that endogenous enzymes will be used to process animal feeds, but this is not relevant to human food where the criteria are somewhat different.

PHYTOHAEMAGGLUTININS

These substances, also known as lectins, are glycoproteins widely present in legumes. Since they have the very useful property of selectively combining with hexose residues on glycoproteins, they have been much studied. In particular they are used in blood typing, since they can distinguish the different carbohydrate residues on the surface of erythrocytes which determine the common blood groups. And herein lies the problem, because when this happens *in vivo* it brings about haemolysis. Although they are lethal on injection, when ingested they seem to severely affect the absorption mechanisms, probably by adhering to the surface of the mucosal cells, and haemolysis may not be so evident. A rise in circulating bile pigments after eating beans might indicate a slight increase in haem degradation by the liver, but implies the absorption through the gut of intact haemagglutinin. It has recently been shown that in rats, kidney-bean lectins are very resistant to proteolysis and do bind to the brush-border cells of the intestine. At least 5% of ingested lectin appears quickly in the bloodstream, apparently intact and still active. However tomato lectins are not absorbed under the same circumstances and it may be that protease resistance is a feature of the more toxic lectins.

The toxicity problem is mostly met with *Phaseolus vulgaris*, the common bean. Matters are complicated because this species exists in hundreds of varieties, which appear to contain haemagglutinins of differing specificity. Lima beans, for example, are specific for blood group A. When this is combined with the individual variaton associated with blood groups, and other cell types, it is clear that some populations may be more at risk than others, and some individuals. Even garden peas contain lectins, at about one-tenth the level in *Phaseolus*. Ricin, the lectin from castor beans is so highly toxic that it requires extreme precaution for laboratory work.

The haemagglutinins are normally inactivated by heat. Cooking practices are only just adequate however and there is no margin for error. It has been reported that the lower boiling point of water at high altitude can be enough to prevent inactivation. There have been numerous reports of illness due to inadequately unfolded haemagglutinins.

There is clearly a case for the development of *Phaseolus* strains with destabilized haemagglutinins either towards temperature or proteases. This is also a factor which will affect the growing use of soy protein in human food. As has already been said in other contexts, if we knew what haemagglutinins did in the seed it might be easier to modify them without deleterious effects on their cultivation.

The lectins are highly conserved metallo-proteins and very widespread which implies that they do have an important role. It is generally believed that they are involved in two functions, both involving the recognition of bacterial or fungal surface carbohydrate. It has been proposed that they are associated with the formation of root nodules by rhizobia, by assisting the initial adherence of the bacteria. The other function is a general defence mechanism against fungi and bacteria. Certainly lectins can bind to fungi: what is not clear is how this prevents invasion. It is just possible that they represent some sort of defence mechanism against animal predators, since some of them are very toxic. The effect on rat intestinal mucosa has already been mentioned. They also affect insect pests. *Phaseolus* has two potential storage pests, the bean weevil (*Acanthoscelides obtectus*) and the related *Callosobruchis maculatus*, which infests cowpeas. *C. maculatus* is killed by the lectin, but *A. obtectus* is much less affected. The gut epithelium binds the lectin in the former but not in the latter, while the extent of proteolysis of the two lectins is similar. This does suggest that the ability of lectins to bind to the surface carbohydrate, which has certainly been useful in experimental work, may also lie at the basis of their biological function. It is difficult to arrive at clear conclusions because the learning ability of the predators, and very complex ecological balances are all involved.

Molecular Biology
Two groups of lectins can be distinguished, where the subunits are all identical and those where there are two classes of subunit. They all contain four subunits with extensive sequence homologies. cDNA from soy and pea lectins have been sequenced and many others by direct amino acid analysis. Wheat lectin has the unusual pyroglutamate as the N-terminal. Each subunit has a hexose-binding site, in some cases two. It should be made clear that the hexose in question can be part of a larger oligosaccharide, and the binding site appears to involve the shape of the oligosaccharide in determining specificity. Some of these are shown in Fig. 6.8.

Soy contains at least two related genes for lectins, only one of which provides the seed version. There are lectin-deficient lines, with non-transcribable genes. mRNA production seems to follow the same pattern as the other seed storage proteins.

Where two different subunit types are present, as in ricin, there is evidence that they are synthesized as a single chain, followed by fission later, actually in the protein body. This is a common pattern in seed globulins (see Chapter 3). There may also be a transient attachment of N-acetylglucosamine between the endoplasmic reticulum—the site of synthesis—and the protein body.

One of the interesting features of the bonding between soy agglutinin and N-acetylgalactosamine is that it is much faster than would be expected on a simple diffusion basis. This, which is reminiscent of the way in which enzymes interact with matrices much more effectively than would be anticipated, is a reflection of 'goodness of fit' between the hexose and the binding site.

There is now a steadily increasing body of knowledge about the amino acid

166   Biotechnology in the Food Industry

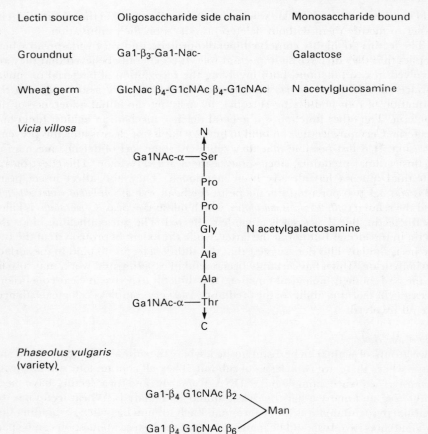

Fig. 6.8 Some oligosaccharides and monosaccharides bound by lectins found in food plants.

sequences, which opens up the possibility of modification. There is much less information about the structure of the oligosaccharide binding sites, which are at least as important in the functional sense. It is not possible at present to forecast the effect of modification on the nutritional value of the bean.

The enzyme processes of the future which may replace fermentation by microorganisms with controlled mixtures of enzymes will have to be designed to eliminate the active haemagglutinins. That they do so is one of the major advantages of current processes.

Population Polymorphisms and Food

Man is obviously an omnivore, but what was he some 2 million years ago before settled agriculture had been thought of? The archaeological evidence, such as it is, is that he was also an omnivore. Most notably, when he obtained food mainly by hunting, it would have been relatively high in fat and protein, and poorer in carbohydrate compared with the diets of the present day. Some 9000 years ago mankind invented agriculture, considerably increased the intake of carbohydrate, and increased in numbers so that hunting was no longer possible as the primary food supply. It is notable that as soon as the standard of living permits it, the level of protein in the diet starts to rise, something that seems to be quite unrelated to the general cultural background, or the established pattern of the diet. Yudkin, who made these points in an article published in 1969 would probably argue that the population is drifting back towards its ancestral preferred diet, whenever it gets the chance. It has also been noted that the fatty acid composition of 'wild' animals as opposed to domesticated ones is different and, it has been claimed, nutritionally superior. We are concerned with a different point here, though it is equally controversial. This is the possibility that when agriculture was invented, it brought about a marked shift in human diets, to which we have not yet fully adapted. We remarked in Chapter 4, that the amino acid composition of cereal protein, easily our largest source of protein, was actually short of essential amino acids. Estimates of the daily requirements for these have fluctuated over the years, and are now thought to be lower than the estimates of a few years ago, but growing children in some parts of the world have some difficulty in obtaining a fully adequate balance of amino acids (Fig. 6.9). Indeed the marked increase in average height, which has taken place in the richer parts of the world, and can still be seen happening in some developing countries, is connected in some way with improving nutritional status in childhood.

Yudkin went further and suggested that if people's food preferences were examined, without any kind of economic restraint, then it would probably reflect their ancestral and thus optimal diet. He is here making the assumption that man, in much the same way as other omnivores has some sort of in-built ability to choose an adequate diet. This is an assumption which at the least requires examination, and proof is not likely to be easy to find. However, meat, fruit and a small amount of root carbohydrate, are probably near the pre-agricultural diet, and are close to modern preferences where they are unmodified by cultural pressures. That is, no-one would expect an officially vegetarian society to show a marked preference for meat, but if we look at the world as a whole, and the amazing variety of food taboos and imperatives, which largely cancel each other out, we can discern an underlying pattern of preference.

Much nearer our own times, the industrialized world has seen another revolution in its food patterns, again associated with an enormous increase in population which makes reversion to an older style of production and consumption impossible. The earlier revolution has not yet been fully adapted for: reaction to cereal protein and dairy products are the commonest in those populations that have been studied, largely Western and European in origin. This suggests that these novelties of 9000 years ago are still capable of causing problems in some populations. Do we have equal problems arising from the changes of the last 200 years, and as the process of change continues, are we likely to have more? This

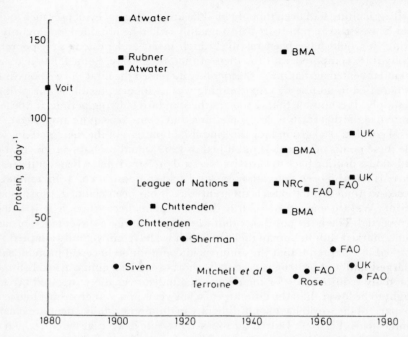

Fig. 6.9 The recommended protein intake for a 65 kg man and estimates of the minimum protein to retain positive nitrogen balance, as they have varied over the years. The names are of individual authors of published recommendations or BMA (British Medical Association), FAO (Food and Agriculture Organization), NRC (National Research Council, USA) and UK (Department of Health). (From P.R. Payne (1978), *Plant Proteins*. Ed. G. Norton. Butterworths, London.)

possibility is the reason for introducing this topic into a book about biotechnology in the food industry because the greatly enhanced feasibilities inherent in the application of the techniques may accelerate the processes of change. Adaptation requires time. In addition we now have a great deal more information about the variability of populations, itself a branch of molecular biology, and may be able to do something to anticipate difficulties if any should emerge.

It has been argued that food manufacturers cannot be expected to take account of the problems of a tiny minority of potential customers. For example, most people are aware of foods that 'do not agree with them' and quickly learn to avoid them. The question is obviously a controversial one, and will be for a long time yet. However, aspartame is now sold with a warning to phenylketonurics printed on the container, and they are approximately 1 in 15 000 of the European populations. (It is improbable that phenylketonurics will be unaware of their condition. In the UK all babies are tested at birth.) If this warning principle is followed then there may be many more: those in favour argue that this is needed

Food, Populations and Quality Control   169

because as processing increases, or now even the raw materials may alter as a result of gene transfer, information on content is not easily available to potential consumers. Labelling standards are inadequate even when they know that they should avoid certain substances. Most food manufacturers are now aware that labels will have to be more informative, and are awaiting legislation. As will be seen another problem is the great difficulty of testing foods for some potential hazards. Responsible food manufacturers carry out quality control to assure themselves that their products are adequate, but can only do what is feasible. In the following sections we will review a few of the polymorphisms that are known to occur and the way in which biotechnology may be able to help.

COELIAC DISEASE

This is one of the commonest conditions that might be due to the adoption of agriculture in the distant past. It is caused quite simply by ingesting gluten from any one of wheat, barley, rye, buckwheat or derived products such as malt. Cooked and processed foodstuffs containing wheat flour are equally effective in causing the condition. It has even been claimed that tryptic digests can cause it: this seems logical since the wheat protein would be subjected to at least partial degradation by the normal digestive enzymes, which are present. The disease is characterized by marked changes in the mucosa of the gut, and defective absorption. In many cases there is a protein losing enteropathy which leads to general malnutrition and low levels of circulating serum proteins. This together with loss of water and salts accounts for most of the symptoms. It is often associated with a marked deficiency of lactase, and a relatively high incidence of diabetes mellitus. All the symptoms disappear in a few days on a gluten-free diet.

Despite the relative frequency of the condition it is still unknown precisely which of the gluten proteins is responsible for the mucosal damage. The first suspect would obviously be the lectins, and the hereditary nature of the condition suggests that there is an individual susceptibility to them. The most likely situation is that the surface carbohydrates of the mucosal cells are able to interact with lectins present in the seeds. This may not be the whole story however and it is possible that the lack of a specific peptidase, or even an amino acid absorption mechanism is also involved.

FAVISM

This condition which has many names, the most charming being 'Baghdad Spring Fever', is caused by eating broad beans (*Vicia faba*), but only in some people. In Baghdad they were eaten at a Spring Festival. The condition involves haemolysis, and may if prolonged lead to anaemia. It is because only a sector of the population suffer such ill effects that it has attracted a lot of attention, but it really serves as an example of what may be a more frequent situation.

Figure 6.10 shows a number of components present in the bean, of which divicine is believed to be the major one, which may reduce the level of glutathione (GSH) in the tissues. For some unknown reason, GSH is very abundant in human erythrocytes. When individuals deficient in glucose-6-phosphate dehydrogenase

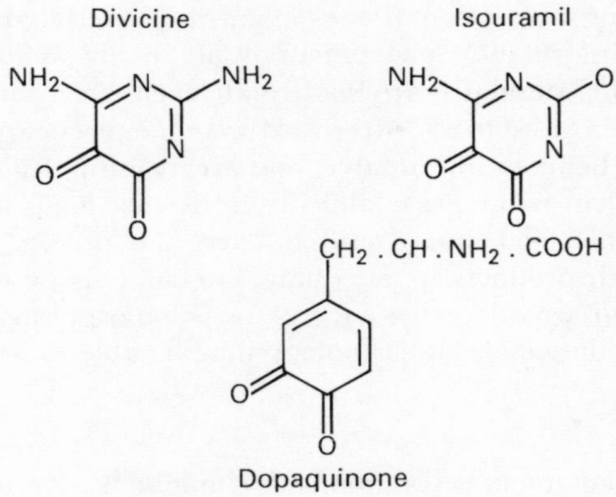

Fig. 6.10 Compounds found in *Vicia faba*, believed to oxidize glutathione, and contribute to favism. They are all shown in their oxidized form, but being quinonoid can act as general oxidizing agents.

activity also suffer a loss of GSH, the red cells lyse. (Note that we refer to a lack of activity rather than a lack of the enzyme. Point mutations lead to the alteration of a single amino acid in the chain, and the enzyme may well be present, but inactive. It is inaccurate in such circumstances to refer to a lack of the enzyme, though this is often done.) A difficulty with this general account is that these substances also occur in other beans, including peas, but have not usually been associated with favism. There have been claims that even the inhaled pollen of the beans can precipitate an attack, so it is not likely to be due to a massive ingestion of haemagglutinins. It seems certain however that it is a number of effects coming together, with one of them, the level of glucose-6-phosphate dehydrogenase under genetic control. There is no true animal model for favism, though hens are also affected by divicine in the diet. Heating the bean does not eliminate the effect. There is actually a fairly close correlation between the presence of malaria in a population and the dehydrogenase deficiency, and thus this may give some selective advantage against malaria. It is known that the malarial parasite, which infests erythrocytes, derives much of its cysteine requirements from the GSH.

For the food manufacturer, who is selling *Vicia faba* for consumption, and always has done, the possible presence of such individuals among his customers presents a dilemma. This becomes even more acute when he may be incorporating a raw material derived from *V. faba* into some manufactured food. When something has been used as an article of food for centuries the pressure is less than when it appears to be a new introduction. The matter remains unresolved, and for the moment must depend on individuals knowing their own susceptibilities, and accurate labelling. At least the effects can now be linked more or less directly with their cause. The incidence varies considerably between populations. In Europe it is common all round the Mediterranean coasts and it has not so far been significant

Food, Populations and Quality Control

in areas where manufactured foods are widely used. As food manufacturing spreads, and populations move, favism has a good chance of being the second condition which will require mandatory labelling of the presence of a potentially harmful component. There is some evidence that vitamin E supplementation may protect against divicine, but this comes only from studies with rats.

HEXOSE INTOLERANCES

Hexoses are absorbed from the gut by active processes, and problems may occur either because of a deficiency of digestive enzymes, or in absorption, or in subsequent metabolism. We have already referred to the inability of most Amerindians, Chinese and Arab adults to utilize lactose. Lactase activity, as might be expected is high in small children, and tends to diminish in adults to about 10% of the initial level. In these populations it almost disappears, leading to all the effects of unabsorbed carbohydrate in the gut if milk is consumed. There are rare inherited conditions leading to failure of the absorption mechanisms for all or some hexoses. Galactosaemia is first seen when children fail to make progress on a normal milk diet. Avoidance of galactose brings about a marked improvement, but otherwise severe damage ensues. The reactions of galactose are shown in Fig. 6.11, and the lesion is usually in galactosyl-1-phosphate uridyl transferase, which has no activity. Galactose-1-phosphate builds up in the tissues and provokes a hypoglucosaemia, which leads to generalized damage.

In fructose intolerance, the enzyme responsible is aldolase (EC 4.1.2.13), and most often it is sucrose which must be removed from the diet, as well as fructose itself. An adult can ingest up to 100 g of fructose daily from fruit and confectionery sources. In most cases it is the specific aldolase of the liver that is lacking and the muscle form is intact. Figure 6.12 shows the metabolic pathways. The lesion leads to a build-up of fructose-1-phosphate. In a rare condition where fructokinase is lacking, the build-up is of fructose itself, which appears to have no ill effects.

The incidence of fructose intolerance is about 1 in 20 000 in Europe, and because it is relatively rare it is sometimes not diagnosed correctly. It usually first appears on weaning, since lactose causes no problems but changing to sucrose leads to the first exposure. Weaning is a very significant event in the development of a child, and sadly some unrecognized sufferers have been subjected to entirely unnecessary psychotherapy, along with their mothers! It is not easy to confirm aldolase deficiency—liver biopsy should not be undertaken lightly—but the gene has been located and a direct DNA probe should shortly be available. This will also permit the identification of carriers at about 1 in 140 of the population. For no obvious reason, except perhaps abstention from sucrose, fructose intolerance is associated with an absence of dental caries.

PROTEINASE DEFICIENCIES

Trypsinogen deficiency is a condition in which nearly all the pancreatic proteinase activities are absent. This is because in the absence of trypsin, chymotrypsinogen and procarboxypeptidase are not converted to the active forms. There is therefore an almost complete failure of protein digestion. This is of course consistent with the

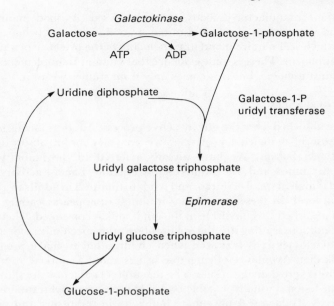

Fig. 6.11 The enzymes involved in galactosaemia. The galactosyl-1-phosphate uridyl transferase has no activity, so that galactose-1-phosphate accumulates.

effects of trypsin inhibitors already discussed in Chapter 3. The condition is first recognized in infancy, naturally, and the child can usually be saved by administering hydrolysed protein or even mixtures of amino acids. There is a similar hereditary condition where enterokinase, the enzyme that converts trypsinogen to trypsin, is absent with the same consequences for protein digestion.

LIPASE DEFICIT

Many individuals are known with low levels of lipase activity, and as would be expected they have defective fat absorption, more acute but otherwise similar to the effects of low bile production.

CONCLUSION

This brief discussion covers only a small number of the hundreds of enzyme defects which have been recognized in the human population. Many of them lead to severe metabolic defects, and very often death. Others have no obvious clinical effects, and may be hard to find. There are a few where the effects are significant but are not life-threatening, and some of these interact directly with the diet. They may require very substantial modification to the average diet, or they may be unaffected by the local one but at risk from substances present in other diets. As diets change, and novel components are introduced the possibility exists that

Food, Populations and Quality Control

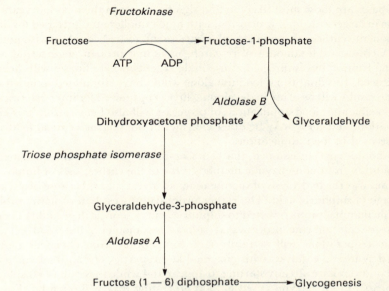

Fig. 6.12 Enzymes associated with fructose intolerance: the reactions catalysed by aldolase. In most cases fructose-1-phosphate accumulates because of an absence of aldolase B activity in the liver. Aldolase A in muscle still functions, though both forms can catalyse both reactions.

hitherto unrecognized enzyme deficiencies may suddenly be revealed. An increase in our knowledge of population variation may help to avert this, which otherwise seems impossible to avoid. Completely new foods should be introduced with some caution. Happily the evidence from the fairly numerous introductions of the last two hundred years suggests that the numbers involved may not be very large, and the effects not too severe.

General Conclusions and Future Prospects

Novelty in the food industry is almost always a matter of finding and using new ingredients, or of changing a process. In both cases these lead to a product similar to an existing one. This is so much the case that even those embodiments of novelty—patents—are classified as either 'composition' or 'process' in nature. Really new processed foods appear much more rarely on supermarket shelves, and even when they do are often transplants from other countries. Within this innovatory framework there are three major themes in food biotechnology.

The first is the application of specific enzyme activities to make raw materials or ingredients. The most successful is undoubtedly the transformation of starch to maltose, glucose and fructose by a variety of enzymes. Of enzymes not currently in use, lipases are those most likely to find significant use in the next few years. This will be in large-scale lipid processing, with much smaller amounts used for flavour component synthesis. Lipases also illustrate an interesting coupling with non-food use. A considerable amount of research, including protein engineering is now nearing fruition but is directed at use in detergents, where lipases will shortly be introduced. It is doubtful if food use alone would have stimulated so much work, but the results will be of use in developing lipases optimized for inter-esterification. It is possible that food applications of proteases will also benefit from this kind of association, and there may be other enzymes in the pharmaceutical field which can be used in food applications.

It should not be assumed that enzyme-based processes are immune from competition from non-enzymic methods. One of the earliest uses of immobilized enzymes was the hydrolysis of sucrose to make invert sugar, because of a wartime shortage of sulphuric acid. The rather primitive process had so many problems that the manufacturers reverted to sulphuric acid as soon as they could. There are no large-scale enzyme processes at risk at the moment, but small-scale fine chemical operations will certainly face competition from chemical synthesis. Indeed synthetic catalysts with enzyme-like properties of chiral specificity and mild conditions are already starting to appear and while they will not displace the far cheaper crude enzymes used in food processing they will often be chosen for chemical synthesis in preference to enzymes of natural origin.

It would also be dangerous to assume that the climate of opinion that gives 'natural' processes based on enzymes a marketing advantage over non-enzyme processes will last indefinitely. This fashionable view is not universal and could disappear as quickly as it arrived to be replaced by something equally irrational. The present situation in the sweetener market, where sucrose, enzyme-produced fructose, part microbial aspartame, and purely synthetic saccharin and acesulfame, all coexist and compete is a more representative example of the way food ingredients are likely to develop.

The second major theme is a steady increase in our knowledge about well-established fermentation processes. There is a growing number of examples where the controlled use of extra enzymes to aid or modify fermentation has found commercial application. This is most highly developed in cheese manufacture. Ultimately many such processes will eliminate the microorganisms, at least in part using instead cocktails of enzymes. One cheese is now made in this way in the USA but it will be many years before a significant proportion of world cheese production goes over to this kind of process. Chymosin from yeast is likely to be widely used first.

Soy beans and other legume-based products will almost certainly attract much interest in the controlled use of enzymes. There is clearly scope for the development of proteases targeted at undesirables such as trypsin inhibitors and lectins. This would do much to make beans amenable to a modern process technology. However the use of enzymes as extraction aids, for lipids or proteins will probably be the first application in legume processing. Looking further ahead, controlled enzymic processing will be developed to make products equivalent to the oriental textured soy foods. Detoxification methods based on enzymes could clearly be improved and replace some of the more hazardous traditional ones. Whether this happens will depend on the economies of the developing world which may have other priorities.

The third theme is the modification of organisms to produce new raw materials. The most striking example of this in the last few years has been the sharp increase of rape-seed oil output. In fact classical plant breeding methods were used to propagate a low erucic acid variety, and this together with a favourable political climate was responsible.

The example discussed in Chapter 1 was the first demonstration that lipid synthesis can be modified by transferring enzymes, but is clearly some way away from commercial use in oil seeds. Another attractive possibility in this field is to alter the balance of starch and lipid as storage materials in seeds. A high-lipid low-starch pea would be interesting, and probably feasible. In the longer term, developments of this kind will have far-reaching effects on food processors, but they are long term and although much work is being done there are no immediate applications in sight.

There are other branches of biotechnology and other points of view. In this book we have more or less consistently explored the view that would be taken by food processors. The producers of enzymes will have different interests. Few of them are part of the food industry. For the most part they started production to meet the large-scale demand caused by the introduction of enzyme detergents (at least half) and the starch saccharification processes (about one-third of total enzyme production), and used expertise in large-scale fermentation gained mostly in antibiotic manufacture. Their commercial interests lie not only in increased use but in adding value to their enzymes. Rather than sell a crude concentrate of a culture medium, immobilized enzymes on support media suitable for use in columns are now made. The same producers are also involved in making the highly purified enzymes needed for some pharmaceuticals, and no doubt could develop supplies of these if they should ever be required in the food industry. Whether this will extend to protein engineering remains to be seen.

One possible way ahead is for very large food companies to do their own genetic manipulation, and produce the transgenic organisms, followed by licensed manufacture by the enzyme producers. At present it is not at all clear where most of this work will come to rest. A large complex industry like food will develop by small incremental steps and it is unlikely that there will be any spectacular novelties. Over the last 200 years the diet of developed countries has altered a great deal and this is a continuing trend with which biotechnology in its various manifestations has always been associated. The trend is as it has been for a long time to widen the variety of foods available and by preservation to make them available at all times. Increase in variety in the past has been obtained by introducing food from one part of the world to another. There are some exceptions. Margarine did not arrive in this way, and the most recent 'new food' in the UK is thought to be breakfast cereals, still unknown in the rest of Europe. It is quite impossible to predict what the next one will be, or whether biotechnology will have any part to play in it, though new fermented foods and some new fruits are clearly on the horizon.

It is not surprising, but it is unfortunate, that the food industry is most highly developed in the parts of the world that have a food surplus. The industry is very good at discovering and using new raw materials, or old ones in novel and more efficient ways. In the next 20 years or so as food production per capita in many parts of the world follows existing trends and turns downwards this ability of the food industry will be needed. Despite the probability that it will be strongest where it will be needed least, biotechnology will make a significant contribution. In a perfect world, biotechnology would be aimed at making marginal food sources in areas of food shortage a good deal safer than they are now. There are many potential food plants that need relatively little development to be made less hazardous.

Whatever happens, biotechnology, the application of enzymes in a controlled way, will have a part to play.

Further Reading

Chapter 1

The following books, reviews and a few original papers are relevant to this chapter. They are meant to provide a view of the subject in depth. They are not designed to cover the most recent developments: this is adequately done by a plethora of annual reviews.

Butler, L.O., Harwood, C. and Moseley, B.E. (Eds) (1989). *Genetic Transformation and Expression*. Intercept, Wimbourne, Dorset.
Dean, P.D.G., Johnson, W.S. and Middle, F.A. (1985). *Affinity Chromatography: A Practical Approach*. IRL Press, Oxford.
Edens, L. and Van der Wel, H. (1985). 'Microbial synthesis of the sweet tasting plant protein thaumatin', *Trends in Biotechnology* **3**, pp. 61–64.
Esser, K. and Kamper, J. (1988). 'Transformation systems in yeasts. Fundamentals and applications in biotechnology', *Process Biochem.* **23**, pp. 36–41.
Gacesa, P. and Hubble, J. (1987). *Enzyme Technology*. Open University Press, Milton Keynes.
Hilditch, T.P. (1956). *Chemical Constitution of Natural Fats*. Chapman and Hall, London.
Hitchcock, C. and Nichols, B. (1971). *Plant Lipid Biochemistry*. Academic Press, London.
Hoffman, L.M., Donaldson, D.D. and Herman, E.M. (1988). 'A modified storage protein is synthesized, processed and degraded in the seeds of transgenic plants', *Plant Molec. Biol.* **11**, pp. 717–729.
Kershavarz, E., Hoare, M. and Dunnill, P. (1987). 'Biochemical engineering aspects of cell disruption', in *Separations for Biotechnology*. Eds M. Verrall and M. Hudson. Ellis Horwood, Chichester.
Lathe, R. (1985). 'Synthetic oligonucleotide probes deduced from amino acid sequence data', *J. Molec. Biol.* **183**, pp. 1–12.
Lynen, F. (1980). 'On the structure of the fatty acid synthetase of yeast', *Eur. J. Biochem.* **112**, pp. 434–442.

Neidleman, S.L. (1987). 'Effects of temperature on lipid unsaturation', in *Biotechnology and Genetic Engineering Reviews*, vol. **5**. Intercept, Wimborne, Dorset.

Ness, W. R. and Ness, W.D. (1980). *Lipids in Evolution*. Plenum, New York.

Shah, D.M., Tumer, N.E., Fischoff, D.A., Horsch, R.B., Rogers, S.G., Fraley, R.T. and Jaworski, E.G. (1987). 'The introduction and expression of foreign genes in plants', in *Biotechnology and Genetic Engineering Reviews*, vol. **5**. Intercept, Wimborne, Dorset.

Shewry, P.R., Kreis, M., Burrell, M.M. and Miflin, B.J. (1987). 'Improvement of the processing properties of British crops by genetic engineering', in *Food Biotechnology*. Eds R.D. King and P.J. Cheetham. Elsevier Applied Science, London.

Scopes, R.K. (1982). *Protein Purification. Principles and Practice*. Springer-Verlag, New York.

Tombs, M.P., Cooke, K.B., Burston, D. and Maclagan, N.F. (1961). 'The chromatography of normal serum proteins', *Biochem. J*. **80**, p. 284.

Volpe, J.J. and Vagelos, P.R. (1973). 'Saturated fatty acid biosynthesis and its regulation', *Ann. Rev. Biochem*. **42**, pp. 21–61.

Webb, E.C. (Ed.) (1984). *Enzyme Nomenclature*. Academic Press, London.

Chapter 2

Bucke, C. (1983). 'There is more to sweeteners than sweetness', *Trends Biotechnol*. **1**, pp. 67–71.

Evans, C.T. *et al*. (1987). 'A novel efficient biotransformation for the production of L-phenylalanine', *Bio-Technology* **5**, pp. 818–822.

Higginbotham, J. (1983). 'Recent developments in non-nutritive sweeteners', in *Developments in Sweeteners*, 2nd edn. Eds T.H. Grenby, K.J. Parker and M.G. Lindley. Elsevier Applied Science, Barking.

Hamilton, B.K., Hsiaio, H.Y., Swann, W.E. Anderson, D.M. and Delente, J.J. (1985). 'Manufacture of L-amino acids with bioreactors', *Trends Biotechnol*. **3**, pp. 64–68.

Harada, T. (1984). 'Isoamylase and its industrial significance in the production of sugars from starch', in *Biotechnology and Genetic Engineering Reviews*, vol. **1**. Intercept, Wimborne, Dorset.

McAllister, R.V. (1980). *Immobilised Enzymes for Food Processing*. CRC Press, Florida.

Schneider, H., Johnson, K. and Mackenzie, C. (1989). 'Enzymic debranching of polysaccharides: effects on polymer properties and on enzymic hydrolysis of the main chain', in *Biotechnology and Genetic Engineering Reviews*, vol. **7**. Intercept, Wimborne, Dorset.

Tombs, M.P. (1985). 'Stability of enzymes', *J. Appl. Biochem*. **7**, pp. 3–24.

Tomazic, S.J. and Klibanov, A.M. (1988). 'Mechanism of irreversible thermal inactivation of *Bacillus* amylases', *J. Biol. Chem*. **263**, pp. 3086–3091.

Chapter 3

Adler-Nissen, J. (1986). *Enzymic Hydrolysis of Food Proteins*. Elsevier Applied Science, Barking.

Bajaj, M. and Blundell, T. (1984). 'Evolution and the tertiary structure of proteins', *Ann. Rev. Biophys. Bioeng*. **13**, pp. 453–492.

Bergquist, P.L., Love, D.R., Croft, J.E., Streiff, M.B., Daniel, R.M. and Morgan, W.H. (1987). 'Genetics and potential biotechnological applications of thermophilic and

Further Reading

extremely thermophilic micro-organisms', in *Biotechnology and Genetic Engineering Reviews*, vol. **5**. Intercept, Wimborne, Dorset.

Bigelow, C.C. (1967). 'On the average hydrophobicity of proteins and the relation between it and protein structure', *J. Theoret. Biol.* **16**, pp. 187–211.

Bond, J. and Butler, P.E. (1987). 'Intra-cellular proteases', *Ann. Rev. Biochem.* **56**, pp. 333–364.

Bone, R., Silen, J.L. and Agard, D.A. (1989). 'Structural plasticity broadens the specificity of an engineered protease', *Nature*, **339**, pp. 191–195.

Campbell-Platt, G. (1987). *Fermented Foods of the World*. Butterworths, London.

Clark, A.H., Judge, F., Richards, J., Stubbs, J. and Sugget, A. (1981). 'Electron microscopy of network structures in thermally induced globular protein gels', *Int. J. Peptide Protein Res.* **17**, pp. 380–392.

Cowan, D., Daniel, R. and Morgan, H. (1985). 'Thermophilic proteases properties and potential applications', *Trends Biotechnol.* **3**, pp. 68–72.

Danilenko, A.N., Grozov, E., Bikbov, T., Grinberg, V. and Tolstoguzov, V. (1985). *Int. J. Biol. Macromol.* **7**, 109–115.

Daussant, J., Mosse, J. and Vaughan, J. (Eds) (1983). *Seed Proteins*. Academic Press, London.

Dickinson, E. and Stainsby, G. (1982). *Colloids in Food*. Applied Science Publishers, Barking.

Dickinson, C.D., Floener, L., Evans, R. and Nielsen, N. (1987). 'Engineering of soybean seed storage proteins', *Fed. Proc.* **46**, p. 2023.

Honig, D. and Wolf, W. (1987). 'Mineral and phytate content and solubility of soybean protein isolates', *J. Agric. Food Chem.* **35**, pp. 583–588.

Lawrie, R. (Ed.) (1969). *Proteins as Human Food*. Butterworths, London.

Matheis, G. and Whitaker, J.R. (1987). 'Enzymatic cross linking of proteins applicable to foods', *J. Food Biochem.* **11**, pp. 309–327.

Norton, G. (Ed.) (1976). *Plant Proteins*. Butterworths, London.

Ogston, A.G. (1958). 'The spaces in a uniform random suspension of fibres', *Trans. Farad. Soc.* **54**, pp. 1754–1757.

Richards, F.M. (1977). 'Areas, volumes, packing and protein structure', *Ann. Rev. Biophys. Bioeng.* **6**, pp. 151–176.

Schellman, J., Lindorfer, M., Hawkes, R. and Grutter, M. (1981). 'Mutations and protein stability', *Biopolymers* **20**, pp. 1989–1999.

Stellwagen, E. and Wilgus, H. (1978). 'Relationship of the protein thermostability to accessible surface area', *Nature* **275**, pp. 342–343.

Tanford, C. (1973). *The Hydrophobic Effect*. John Wiley, New York.

Tombs, M.P. (1967). 'Protein bodies of the soybean', *Plant Physiol.* **42**, pp. 797–813.

Tombs, M.P. (1974). 'Gelation of globular proteins', *Faraday Disc. Chem. Soc.* **57**, pp. 158–164.

Wright, D.J. (1986). 'The seed globulins', in *Developments in Food Proteins*, vol. **4**. Ed. B. Hudson. Elsevier Applied Science, Barking.

Chapter 4

Berg, H.C. (1983). *Random Walks in Biology*. Princeton University Press, Princeton, N.J.

Blanshard, J.M.V., Frazier P.J. and Galliard, T. (Eds) (1986). *Chemistry and Physics of Baking*. Royal Society of Chemistry, London.

Freedman, R.B. (1984). 'Native disulphide bond formation in protein biosynthesis: evidence for the role of protein disulphide isomerase', *Trends Biochem. Sci.* **4**, p. 438.

Henis, Y., Yaron, R., Lamed, R., Rishpon, J., Sahar, E. and Katchalski Katzir, E. (1988). 'Mobility of enzymes on insoluble substrates: the β-amylase starch gel system', *Biopolymers* **27**, pp. 123–138.
Hough, J.S., Briggs, D., Stevens, R. and Young, T. (1982). *Malting and Brewing Sciences*. Chapman and Hall, London.
Munn, E.A., Feinstein, A. and Greville, G. (1971). 'Isolation and properties of the protein calliphorin', *Biochem. J.* **124**, pp. 367–374.
Peppler, H.J. and Reed, G. (1987). 'Enzymes in food and feed processing', in *Biotechnology*. Eds. H. Rehm and G. Reed, vol. **7a**. VCH, Weinheim, W. Germany.
Russell, G.E. (Ed.) (1988). *Yeast Biotechnology*. Intercept, Wimbourne, Dorset.
Wood, B.J. (1985). *Microbiology of Fermented Foods*. Elsevier Applied Science, Barking.

Chapter 5

Deetz, J. and Rozzell, J.D. (1988). 'Enzyme catalysed reactions in non-aqueous media', *Trends Biotechnol.* **6**, pp. 15–18.
Dickinson, E. (Ed.) (1987). *Food Emulsion and Foams*. Royal Society of Chemistry, London.
Fournet, B., Leroy, Y., Montreuil, J., Decaro, J., Rovery, M., van Kuik, J. and Vleigenthart, J. (1987). 'Primary structure of the glycans of porcine pancreatic lipase', *Eur. J. Biochem.* **170**, pp. 369–371.
Fowler, M.W. (1986). 'Process strategies for plant cell cultures', *Trends Biotechnol.* **4**, pp. 214–219.
Gatfield, I.L. (1988). 'Enzymatic generation of flavour and aroma components', in *Food Biotechnology*, 2nd edn. Eds R.D. King and P.S.J. Cheetham. Elsevier Applied Science, Barking.
Hsiao, H.Y., Walter, J.F., Anderson, D.M. and Hamilton, B.K. (1988). 'Enzymatic production of amino acids', in *Biotechnology and Genetic Engineering Reviews*. Intercept, Wimborne, Dorset.
Hughes, S.G., Overbecke, N., Robinson, S., Pollock, K. and Smeets, F. (1988). 'Messenger RNA from isolated aleurone cells directs the synthesis of an α-galactosidase found in the endosperm during germination of guar (*Cyamopsis tetrogondoloba*) seed', *Plant Molec. Biol.* **11**, pp. 783–789.
Kimura, A. (1986). 'Molecular breeding of yeasts for production of useful compounds; novel methods of transformation and new vector systems', in *Biotechnology and Genetic Engineering Reviews*, vol. **4**. Intercept, Wimborne, Dorset.
Macrae, A. (1983). 'Lipase catalysed interesterification of oils and fats', *J. Amer. Oil. Chem. Soc.* **60**, pp. 291–294.
Macrae, A.R. and Hammond, R.C. (1985). 'Present and future applications of lipases', *Biotechnology and Genetic Engineering Reviews*, vol. **3**, Intercept, Wimborne, Dorset, pp. 193–217.
Milsom, P.E. (1986). 'Organic acids by fermentation', in *Food Biotechnology*, vol. **1**. Eds R.D. King and P.S.J. Cheetham. Elsevier Applied Science, Barking.
Roberts, G. and Tombs, M.P. (1987). 'Preparation of fluorescent derivatives of lipases and their use in fluorescence energy transfer studies in hydrocarbon–water interfaces', *Biochem. Biophys. Acta* **902**, pp. 327–334.
Tombs, M.P. and Blake, G. (1982). 'Stability and inhibition of *Aspergillus* and *Rhizopus* lipases', *Biochem. Biophys. Acta* **700**, pp. 81–89.
Wisdom, R.A., Dunnil, P. and Lilly, M.D. (1985). 'Enzymic interesterification of fats: the

effect of nonlipase material on immobilised enzyme activity', *Enz. Microb. Technol.* **7**, pp. 567–572.

Zaks, A. and Klibanov, A.M. (1984). 'Enzymic catalysis in organic media', *Science* **224**, pp. 1249–1251.

Chapter 6

Anderson, D. and Cuthbertson, W.F.J. (1987). 'Safety testing of novel food products generated by biotechnology', in *Biotechnology and Genetic Engineering Reviews*, vol. **5**. Intercept, Wimbourne, Dorset.

Angharad, M., Gatehouse, R., Shackley, S., Fenton, K., Brydon, J. and Pusztai, A. (1989). 'Mechanism of seed lectin tolerance by a major insect storage pest of *Phaseolus vulgaris*', *J. Sci. Food Agric.* **47**, pp. 269–280.

Collins, W. (Ed.) (1988). *Complementary Immunoassays*. John Wiley, Chichester.

Fenwick, G.R. (1989). 'Bracken—toxic effects and toxic constituents', *J. Sci. Food Agric.* **46**, pp. 147–174.

Fowden, L. (1978). 'Non-protein nitrogen compounds: toxicity and antagonistic action in relation to amino acid and protein synthesis', in *Plant Proteins*. Ed. G. Norton. Butterworths, London.

Gibbs, J.N. and Kahan, J.S. (1986). 'Biotechnology and the food industry: leaping the regulatory hurdles', *Bio-Technology* **4**, pp. 199–205.

Greenwood, D.J. (1989). 'Plant nutrition and human welfare: the world scene. The 1988 Shell lecture', *J. Sci. Food Agric.* **48**, pp. 387–410.

Hanssen, M. (1984). *E is for Additives*. Thorsons, Wellingborough, UK.

Harris, H. (1975). *The Principles of Human Biochemical Genetics*. North-Holland, Oxford.

Krogh, P. (Ed.) (1987). *Mycotoxins in Food*. Academic Press, London.

Liener, I.E. (1987). 'Detoxifying enzymes', in *Food Biotechnology*, vol. **1**. Eds R.D. King and P.S.J. Cheetham. Elsevier Applied Science, Barking.

Lis, H. and Sharon, N. (1986). 'Lectins as molecules and as tools', *Ann. Rev. Biochem.* **55**, pp. 35–67.

Mattiasson, B. (1988). 'Bioaffinity methods of analysis applied in the food area', in *Food Biotechnology*, vol. **2**. Eds R.D. King and P.S.J. Cheetham. Elsevier Applied Science, Barking.

Morris, B.A. and Clifford, M.N. (Eds) (1985). *Immunoassays in Food Analysis*. Elsevier, London.

Robinson, D. (1987). *Food Biochemistry and Nutritional Value*. Longman Scientific, Harlow, Essex.

Ucko, P.J. and Dimbleby, G.W. (Eds) (1969). *The Domestication and Exploitation of Plants and Animals*. Duckworth, London.

Ulitzer, S. and Kuhn, J. (1987). 'Introduction of *lux* genes into bacteria. A new approach for specific determination of bacteria and their antibiotic susceptibility', in *Bioluminescence and Chemiluminescence: New Perspectives*. Eds J. Scholmerich, R. Andresen, A. Kapp, M. Ernst and W. Woods. John Wiley, Chichester.

Velazquez, A. and Burges, H. (Eds) (1984). *Genetic Factors in Nutrition*. Academic Press, New York.

Index

absorption, of intact lectins, 164
Acetobacter, in vinegar, 126
acetolactate decarboxylase, in brewing, 126
acetolactic acid, in beer, 125
acetyl CoA, 9
acetyl CoA carboxylase, 14
acetyl transferase, of yeast, 9
aconitase, in *A. Niger*, 144
ACP (acyl carrier protein), 9, 10, 11
   gene cloned, 16
   varieties in rape seed, 15
activity loss in reactors, 60
actomyosin, 151
   gels, 83
acylation, of proteins, 109–10
adaptation to dietary change, 167, 168
additives, 148
aflatoxin analysis for, 152–4
agar, 140
aggregation, theory of, 88
agriculture, adoption of, 167
Agrobacterium tumefaciens, 45, 141
akee fruit, 163
albumin, 75
   aggregation of, 77, 86, 88
aldolase, 177, 178
   DNA probe for, 177

alginates, 141
allysine, 76, 78
almonds, 158
amino acids
   abbreviations for, 29
   sequencing, 33–4
   toxic, 162
aminopropionitrile, 163
amygdalin, 157
amylase inhibitors, 123
amylases, 52, 118, 123
   thermostability of, 61
amyloglucosidase, 52
amylopectin, 54
amylose, 54
ancestral diet, 167
antibiotic resistance, 156
antibodies
   amounts required to induce, 20
   in analysis, 149
antigens, 20, 150, 151
α_1-antitrypsin, elastase inhibitor, 102
arabinosans, 96, 119, 138
arachin, antibodies to, 150
ascorbic acid, in bread, 116
aspartame, 48, 66
   risk to phenylketonurics, 168
aspartate proteinases, 104

Index

aspartic acid, production methods, 66
Aspergillus flavus, 96, 153
  nidulans, 91
  oryzae, 96
  sojae, 96
Aspergillus niger
  in milk loaves, 119
  lipases in, 128
  pectinases in, 139
Aspergillus parasiticus, 153
astaxanthin, 143

Bacillus subtilis, 102
  thuringiensis, 42
bacteria, detection of, 154
Baghdad Spring Fever, 169
baking, 114–18
barley, 121
bean weevils, 165
biotin, in baking, 119
blackcurrants, imitation, 140
Blighia sapida, 163
bracken, 161
branching enzymes, 121
Brassica chinensis, 159
Brassicas, glucosinolates in, 159, 160
Brassica napus, 159
bread, 114–18
bromelin, 108
butyryl ACP, 11

cabbage, 160
California buckeye, 163
calliphorin, in blow flies, 114
Candida lipolytica lipase, 143
Capsaicin, from cultures, 143
carboxylic acids, in flavours, 143, 144
carboxymethylcellulose, 23
carboxypeptidase
  activity, 79, 100
  cloning, 102
  water inside, 73
Carica papaya, 105
carragenins, 140
carrot rot, organism as pectinase, 139
caseins
  in cheese, 90
  cross linking, 78
  gels, 60, 94, 95
  in mammary gland, 37

cassava, 156
castor beans, 164
caviar, 140
cellulases, 45, 55, 96
cereals, 112–20
chalconine, 159
champagne, 126
charge–charge interactions, 74
cheddar cheese, 91
cheese, manufacture, 89–92
chewing gum, Japanese, 65, 68
Chinese cabbages, 159
chocolate, 131
  paste centres, 53
chromobacter lipase, 130
chymosin, source of, 4, 90
chymotrypsin, 67
  gels of, 90
  stability, 80
cinnamic acid, 67
citric acid, production, 144
citronnelyl butyrate, 142
clonal oil palms, 8
coagulation of proteins, 87, 88
cocoa butter
  composition, 131
  substitutes, 131
coconut
  fatty acids in, 8
  oil, 7
coeliac disease, 169
colipases, 131
column chromatography, 26
condensing enzyme, 11–13
conserved proteins, antibodies to, 151
corn syrups, 54
Corynebacterium manihot, 66, 157
  glutamicum, 145
costing, 5
cotton seed, gossypol in, 160
covalent cross links, 74
cuphea, fatty acids in, 20
cyanogenic glycosides, 156
cyclamates, banning of, 61
cysteine, chemistry of, 75

decalactone, 142
dehydratase, 12
denaturation of proteins, 81
desaturases, 8, 18

detoxification, 162
diaminobutyric acid, 163
diethylaminoethyl cellulose, 23
Dioscoreophyllum cumminsii, 62
disulphide links, 75, 77, 116–18
divicine, 170
DNA
  in analysis, 150
  insertion, 45
  polymerase, 37, 107
  probes, 35
  reverse transcriptase, 37
  sequencing, 37–9
dough, 116–18

Edman degradation, 34
electrophoresis, 29–31, 150
  theory of, 29
electroporation, 45
ELISA antibody assay, 151
emulsifiers, 134–5
emulsin, 154
endoproteinases, 100
enoyl reductase, 12
enzymes
  acting on starch, 52
  availability of, 4
  extraction of, 22
  immobilization of, 59
  isolation of, 20–30
  as natural processors, 2
  nomenclature, 3
  powders, hazards of, 149
  scale of isolation, 20, 21
  stability of activity, 60
  strategies for, 24
epimerase, in seaweed, 141
erucic acid, rape seed lacking, 9, 15
Erwinia carotovora, 139
Eskimos, lack of sucrase, 47
essential amino acids, 114, 167
esters, as flavours, 133, 142
ethyl acetate, 142
exons, 39, 41
exopeptidases, 101

FAS, 9–17
fatty acid desaturase, 8
fatty acid synthetase, 9–17
  structure of, 17

fatty acid synthetic enzymes, specific activities, 14
fatty acids
  chain length control, 9
  variation in sources, 8
favism, 169, 170
feasibility, 3
fermented foods, 94–9
  based on soy, 96
ferrocyanide, 144
fibrin clot, 78
filaments, from proteins, 82
flatulance factor
  with phaseolamin, 123
  in soy, 97
flavours, 142, 143
flax seed, 164
fluorescence recovery after photobleaching, on amylase, 54, 123
fructofuranosidase, 52
fructose intolerance, 177
  fractionation from glucose, 58
fructose syrups, 53
fruit, mono and disaccharide content, 49
fucose, 138
fumaric acid, 66
furfural, in bread, 120

galactosaemia, 177
galactosans, 96
galactose, 49, 128
gari, from cassava, 157
gel filtration, 26
gelatin, gelation of, 82, 83, 87
Gelidium, 140
genetic code, 36
Geotrichum candidum, lipase in, 129
ghatti, 138, 139
gliadin, 115
glucoamylase, 52
gluconic acid, 146
gluconolactone, 146
glucose
  consumption of, 57
  isomerization to fructose, 55, 56
  sources of, 55
  syrups, 53
glucose isomerase, 4
  half-life of, 59
  mammalian, 57

Index

temperature dependence, 59
glucose 6-phosphate dehydrogenase, in erythrocytes, 169
glucuronic acid, 138
glutamate
  production of, 145
  in soy, 97
glutamyllysine, ingestion by rats, 78
glutaraldehyde, 60
glutathione, 75, 116, 169, 170
glutenin, 115
glycinin
  antibodies to, 150
  expressed in petunias, 45
glycolipids, as emulsifiers, 136
glycoproteins, 165
glycosides, toxic, 156
glycosylation of proteins, 44, 110
goats, 89, 92, 161
goitre, 158, 159
Gracilaria, 140
groundnut oil, 7
groundnut protein, quality, 87
guar gum, 140
guluronic acid, 138
gums, 139

haemagglutenins, 164–6
hard wheat, 118
hepatoma, 153
hexose intolerance, 177
hops, 126
hydrogen bonds, 74
hydrophilic amino acids, 73
hydrophobicity, of amino acids, 73
hypoglycin, 163

ideas, source of, 1–3
immobilized enzymes
  matrix preparation, 59
  in reactors, 58, 59
inositol phosphate, 98
insect fat bodies, 114
interesterification of fats, 131–3
introns, 39, 41
inulin, sources of, 53
inulinases, 53
invert sugar, 52
invertase, 52
isoamylase, 52

isolation
  of mRNA, 37
  of proteins, 20–5
isouramil, in *Vicia fava*, 170

Jamaican vomiting syndrome, 163
Jerusalem artichoke, as inulin source, 53
jojoba meal, 160

kangaroos, 149
keratin, 75
ketoacyl fatty acids, 11–15
  Ketoacyl reductase, 12
  Ketoacyl synthase, 12
Kluyveromyces lactis, 43, 91
  marxianus, 53
konjac, 141
kwashiorkor, 154

α-lactalbumin, 79, 92
lactase defficiency, 169
lactate dehydrogenase, chiral specificity, 145
lactic acid production, 145
Lactobacillus
  acidophilus, 160
  bulgaricus, 92, 145
  delbruckii, 144, 145
β-lactoglobulin, 92
  cross linking, 78
  in yoghurt, 93
lactose, 92
lathyrism, 162
Lathyrus odonatus, 162
laver bread, 140
leavening, 119
lecithin, 135
lectins, 164, 165
  absorption mechanism, 164
  cDNA of, 165
  of soy beans, 165
  subunits of, 165
legumin, targetted expression in seeds, 45
lima beans, 158
linamarin, 157
linamarinase, 157
linatine, from flax, 163
lingual lipase, 91
linseed, 159

lipases, 128–31
  in bread, 120
  as flavour enhancers, 133
  inhibition of, 131
  PEG derivatives of, 134
lipid sources, 7
lipoprotein lipase, in milk, 133
lipoxygenase in bread, 118
Listeria monocytogenes, DNA probe for, 156
luciferase, 155
lupins, attack by insects, 157
lux genes, 155
  insertion into bacteria, 156
lysine
  availability in lipases, 134
  role in cross linking, 76–9
lysolecithin, 135
lysozyme, 79
  gels of, 90
  mutants of, 82
lysyl oxidase, 78

M13 phage, 37, 38
Macrocystis pyrifera, 141
maize, as starch source, 54
malarial parasite, need for GSH, 170
malonyl CoA, 9
malonyl transferase, 9
malting, 123, 124
maltose, 49, 55
maltotriose, 55, 124
mammalian FAS, folding of, 16
mandelonitrile, 157
Manihot esculenta, 156
manioc, 156
mannose, 128, 138
mannuronic acid, 138
MCH
  amino acid composition, 29
  function of, 20
  interaction with yeast FAS, 32
  isolation of, 28
  peptides, DNA probes for, 35
  sequence of, 39, 40
  on transformed mouse cells, 42
meat, 71, 97
  proteases in, 104
  species identification, 152
medium chain hydrolase *see* MCH

membrane filtration, in whey processing, 92
menthol, 143
metallothioneine, 111
miracle fruit, 61, 62
miraculin, 62
molasses, 53, 144
monellin
  sequence of, 63
  source of, 61
  stability, 61–6
monoclonal antibodies, 152
  preparation of, 153
monoglycerides, 128, 132, 136
Mucor miehei
  lipase, 131
  protease in cheese, 91, 104
mycotoxins, 152
myelomas, in monoclonals, 153
myrosinase, 159

N-acetylglucosamine, 128, 165
napin, regulator for, 45
nicking
  of FAS, 18
  of lipases, 128, 134

Ogston, gel theory, 89
ovalbumin, 78
  as glycoprotein, 111
  unfolding, 79, 80

palm oil production, 7
palmitoyl ACP, 13
pantotheine, 9
pantothenic acid, 9
papain, 100, 103, 104
  beer haze, 104
  in bread, 117
patents, 46, 173
peas, legumin in tobacco, 45, 159
pectin, 138, 139
pentan-2-ol in bread, 120
pentosanases in baking, 119
pepsin, 100, 101, 104
  affinity ligand, 27
peptide emulsifiers, 108
peptides, undesirable, 108
PER, 102
perms, 75

Index

phages, 154
phaseolamin, 123
Phaseolus vulgaris, 164
  glycosides of, 157
L-phenylalanine
  production in Corynebacterium, 67
phenylalanine ammonia lyase, 66, 67
Phenylketonuria, and aspartame, 68
phenylthiohydantoins, 33
phospholipases, 135
phosphorylation, of proteins, 109
phytase, 98
phytic acid, 98
phytohaemagglutenins, 164–6
pI, isoelectric point, 24, 62, 65, 74
Pisum sativum, 159
plant cell cultures, 143
plant-fatty acid synthetase, 11–15, 17
plantains, 164
plants as hosts, 44
plasmids, 41
  as vectors, 41
plasteins, 75
polyclonal antibodies, 150
polyphenolics, as cross linkers, 78
Polysaccharides as stabilizers, 138
population polymorphism, 166–8
post-transcriptional modification, 39, 45
postsynthetic glycosylation, 44
potatoes, 159
probe construction, 35
prolamins, 115
promoters, 43
prospects, 173–6
protease inhibitors, 23, 100–2
proteases, 5, 100–9
protein bodies, 96, 97, 98, 114, 115
protein gels, 82–9
protein storage, 31
proteinase defficiency, 177
proteins
  in bread making, 116
  disulphide isomerase, 75
  interactions with proteins, 72
  stability, 79
protoplasts, 45
pseudomonas-fluorescent, 149
  putida lipase, 129
ptaquilosides in bracken, 161, 162
Pteridium aquilinum, 161

pullulanase, 52, 55, 123

quail, hemlock eater, 156
quality control, 147

raffinose, in beet sugar, 49
random coils, 79
rape seed
  erucic acid in, 7
  glucosinolates in, 159
  oil production, 7
  spinach ACP in, 45
  transformation of, 46
raw materials, cost of, 2
rebaudiosides, 68–70
relative sweetness, 48
rennin, *see* chymosin
research and development, new ideas, 1
research teams for biotechnology, 6
  time scales, 3, 5
restriction nucleases, 41
reverse micelles, for lipases, 134
Rheological parameters, 71–2
rice, 54, 113
ricin, 164

safflower, 160
saffron, 143
Salmonella
  antibodies against, 156
  detection of, 155
Sangers dideoxy method for DNA
  sequencing, 38
saponins, 2
sarcoplasmic proteins, 151
sausage skins, cross linking in, 78
SDS
  gel electrophoresis, 74
  in seed protein classification, 115
sea weed, polysaccharides from, 140
seed proteins, 115
  classification, 116
seeds as transgenic producers, 42
  fatty acid synthetases, 19
serine proteases, 106–7
sheep, transgenic, 42
shikonin, 143
shuttle vectors, 44
site directed mutagenesis, 106, 107
Smoluchowski, aggregation theory, 88

sodium dodecyl sulphate, 30
solanine, 159
sorbitan monooleate, 137
sorghum, 54, 112, 159
soybeans, 96–7
　protein, 78, 94, 108
spinach ACP, in rape seed, 45
stabilizers, 137–41
stability
　maximum temperature of, 81
　of proteins, 79–82
stachyose, 96
starch gelatinization, 54
starch grains, 53, 54
　in bread, 118, 119
　in brewing, 121
　hydrolysis of, 53, 55
　structure of, 122
steps in biotechnology, 5
Stevia rebaudiana, 68
steviosides, 68–70
Streptomyces thermophilus, 92
subtilisin, 82
　in biscuits, 117
　specificity of, 102
　structure of, 105
sucrose
　consumption of, 47
　present in fruits, 49
　preservative effect, 48
　price of, 47
　synthesis of, 70
sugar
　in beans, 50, 51
　price of, 47
sugarbeet, 49
sugarcane, 47, 49
sunflower oil, 7, 19
sweet proteins, 61–6
sweetness structure relationships, 65
swinging arm theory, 11, 13
Synsepalum dulcifica, 61
synthetic enzymes, 174

tapioca, 157
tempeh, 98
terpenoids, 69, 142
texture, 71
thaumatin
　plantations, 62
　pre and pro forms, 43
　production in yeasts, 42, 43
　protease activity in, 64
　sequence of, 63
　source of, 62
　stability, 66
Thaumatococcus danielli, 62
thermodynamic stability, 80
thermophiles, 81, 109
thermolysin, 109
thiaminase, in bracken, 161
thiocyanate
　as cyanide product, 158
　goitrogenic, 159
tokoroten, 140
tomato sauce, 139
toxic amino acids, 162
　components, 156
tragacanth gum, 140
transfer ACP, 13
transglutaminase, 76–8
triacylglycerols, 19
triazine dyes, 27
triglycerides
　deposition of, 19
　evolution of, 8
　synthesis of, 19
trypsin, 101, 102
trypsin inhibitor, 102, 107–8, 109
　in maize, 42
trypsinogen defficiency, 177
tryptophan synthase stability, 82
turkeys, death of, 152

vegetarian cheese, 91
Vibrio fischeri luciferase, 155
Vicia fava, 162
vinegar manufacture, 126

whey, 92, 110
wool, 75
Worcester sauce, 126

xanthans, 141
xylans, 57
xylose, 57, 138
xylose isomerase
　cobalt content, 57
　use for glucose isomerization, 57

Index

yeast
  in bread making, 119
  FAS, 32
  use of as host organisms, 42–4
yoghurt, 92–4
  composition, 93

zamia, 156–8
Zea mais, 54
zein, 115